REPORTER GENOME

Evolving Earth's Organic Black Box

Copyright © 2020 Paul Rommel Elvira

All rights reserved. No part of this book may be reproduced, or stored in a retrieval system, or transmitted in any form or by any means, electronic, mechanical, photocopying, recording, or otherwise, without express written permission of the publisher.

Design and layout by Melanie B. Elvira. Cover from original image by Denis Degioanni on Unsplash.

Library of Congress Control Number: 2018675309

Printed in the United States of America

For

Yñigo and Yulio

Fr. Bel

CONTENTS

Preface *vii*

PART ONE: Specimen and Medium 1
1 A Primer on Life on Earth 3
2 Panspermia 7
3 Zoo Hypothesis 11
4 Panspermia + Zoo Hypothesis 15
5 Evolution of Early Life 19
6 The Importance of Water 25
7 Evolution Leads to Biodiversity 31

PART TWO: Reporter Genome 39
1 Evolution of Intelligence 41
2 Curiosity Gene 47
3 Evolving the Candidate Species 53
4 Human Biology Reflects Habitat 59
5 Genes Mirror the Environment 63
6 Compendium of Earth and Life Histories 69

PART THREE: Potential Co-conspirators 83
1 A Place for Viruses 85
2 Junk DNA 95
3 Modularity of Biological Systems 101
4 Predictability of Evolution 107

PART FOUR: A New Picture 111
1 Synthesis 113
2 A Sublime Display of Altruism 119
3 Implications 123
References 126

PREFACE

In science, there exist a number of big questions that have been begging for answers for so many years now. However, no other topic fans the imagination of biologists more than the questions regarding our origin. When, where, how and why did life appear on this planet? There is just so much at stake in the search for an explanation for our existence. Well, it seems that we are getting closer to the truth.

NASA scientists are confident of finding alien life in the next decade or so. Among the places where space explorers are expecting to find evidences of life is on the red planet, Mars. One of the renowned theoretical physicists in the world Michio Kaku even made a bold prediction that we are going to hear from an alien civilization within this century. Indeed, for many, finding life outside of Earth is no longer a matter of debate but rather a question of when it is going to happen.

This bright prospect of discovering alien life has boosted the stock of one particular hypothesis dealing with the origin of life on Earth - panspermia. If life outside Earth is possible, then it is also possible that this planet could have been a recipient of the precursors or propagules of organisms that led to the proliferation of biodiversity here.

If Michio Kaku turns out to be correct with his prediction, then directed panspermia, or the suggestion that an alien civilization started the evolution of life on Earth, could become one of the possible scenarios that humanity will have to deal with in the future. In any event, we will probably know a thing or two about our role in this universe with such a discovery.

At present, we can only speculate using the limited information that we have about our own evolution. Ideally, there would be hidden clues in the nature of life on Earth that could help us gain any insight about the likely origin of our earliest ancestors. This might not be much to begin with but it should be better than nothing at all. Any bit of evidence that can be used to weave a workable story should be considered invaluable.

I became interested in the origin of life and evolution as a graduate student in Japan. Back then, I was working with marine macroalgae to learn more about the cytoskeletal dynamics of morphogenesis in these relatively simple cells. One idea that struck me as particularly intriguing was the 'behavior' of rhizoids (cells that are analogous to the roots of plants) when presented with various types of substrata.

Algal rhizoids display a variety of responses to surfaces with different physicochemical characteristics. I thought, these simple anchoring implements of algae are amazing because the cells are capable of responding to any kind of material that they encounter in their environment. If the substratum is soft, then the rhizoids would remain in lance-like form and pierce through the medium until the tips branch out thus effectively holding onto the material like a typical plant root. If the surface is hard and smooth, then the rhizoid would flatten and become fixed firmly using a glue-like substance. In between these two mechanisms, a range of distinct reactions of rhizoids are also observed. It seems as if rhizoids know a lot about the environment and must keep a record of different substrata that they have encountered before for use in the event that they encounter the same conditions again. This simplistic view of the cell provided the inspiration for the concept that organisms are potential organic recorders of the environment that can be retrieved and reverse-engineered or reviewed like a black box commonly found in aircraft.

Another inspiration for exploring this book's topic arose from my stint as a postdoctoral research fellow in Singapore where I was involved in investigating the evolution of multicelullarity using freshwater algae. During that time, the concept of modularity in biological systems became clear to me as a ubiquitous phenomenon among organisms. It was also hard to escape the fact that almost all organisms, especially microscopic ones, are quite adept at co-opting modular units such as genes and cellular structures for new purposes that aided survival and enabled evolution.

This book was written to offer a new perspective on evolution under the directed panspermia schema. The challenge here is to view the events on Earth through the eyes of alien evolution researchers. Why? Because in this manner, we can make some sense of several conspicuous features of evolution of life on Earth such as the apparent exclusion of viruses in the tree of life, the prevalence of noncoding or 'junk DNA,' the modularity of biological systems and the predictability of evolution.

In *Part One*, the topic is introduced and the context of this book is laid down by discussing the ramifications of the primordial soup theory, panspermia, zoo hypothesis, the characteristics of life, the importance of water to organisms and the proliferation of biodiversity on Earth. Each of these subjects is essential in reviewing how evolution works and, more importantly, in providing the context for the succeeding concepts in the remaining parts of the book.

In *Part Two*, the mechanisms that led to the emergence of humans as the 'pinnacle' of evolution are discussed through refreshing views on the evolution of intelligence, the relevance of curiosity and the role of gene-environment interaction on the human genome. Chapter by chapter, the idea behind the making of a candidate organism, as a reporter species capable of representing the evolution of life on Earth while illustrating the transformations that this planet underwent over billions of years, gradually becomes evident in this part.

In *Part Three*, the supporting notions for the message of this book are provided using insights on the possible position of viruses in the tree of life as the promoter of genetic change, the potential use of 'junk DNA' for creating new genes, the modularity of biological systems as a promising default strategy for evolving life and the predictability of evolution. These chapters comprise some of the biological principles that can help with fostering the rise of a reporter species - concomitantly serving as the rationale for this thought exercise.

Finally, in *Part Four*, the various narratives in the previous chapters are dovetailed into one coherent story. A takeaway from this synthesis is that if an alien civilization chooses to replay the tape of life on a planet like Earth, the phenomena mentioned in this book could likely provide the justification as well as the theoretical questions for such an endeavor. Conversely, if humans are going to try rerunning evolution for the sake of understanding our origin, then we could pick an Earth-like planet to test the same postulates contained in this book.

It is easy to mistake the thesis forwarded here to be nothing more than an expansion of the directed panspermian hypothesis. This is understandable given the aforementioned assertions but I would rather view this piece as a revelation of the potential of organic life forms to become the medium, message and mechanism- all at the same time - for recording life histories on habitable planets. In a way, this book also attempts to take off where the *selfish gene* theory concludes: eventually, genes (or genomes) have no other way but to be altruistic if life on Earth decides to thrive on other planets as well. A reporter or representative species could make a crucial sacrifice by venturing out to pave the way for other Earth-based species in another home in order to preserve our organic heritage.

PART ONE:
Specimen and Medium

CHAPTER 1

A Primer on Life on Earth

The origin of life is "one of the great unsolved mysteries of science."

– Francis Crick
Life Itself: Its Origin and Nature, 1982

Panspermia posits that life exists in two or more parts of the universe and that life on Earth could have been delivered by comets or by other space objects such as asteroids, meteoroids, planetoids and even space dust. The idea is that Earth, previously pristine, was contaminated by the arrival of aforementioned objects. Interestingly, visits from aliens and their spacecraft are also seen as a possible source of contamination.

Panspermia has been proposed in many forms, particularly in terms of how life reached our planet. Some suggest that the introduction of living organisms or building blocks of life from another planet, star or galaxy was accidental while some hypothesize that advanced civilizations could have deliberately inoculated Earth to start and evolve life on this planet. The manner or purpose of inoculating Earth with life or its starting materials is irrelevant. What is important is that there is an alternative hypothesis on how life came to exist on Earth in addition to the widely accepted idea of abiogenesis.

Abiogenesis suggests that life arose from non-living materials such as basic organic compounds. From simple biochemistry, simple microorganisms took form that gradually evolved into more complex ones such as those found on Earth today. Either way, no one knows for sure what really happened a few billion years ago. Experiments could simulate the conditions on Earth that were favorable for the emergence of organic compounds and the early evolution of life forms during those times but these will not point to a definite origin of life. It is all guesswork from this point on. Since no one can say for certain what really happened, it is thus open for all to hypothesize what could have happened. While at it, anyone can also try to suggest the 'why' of it.

If we chose to believe in abiogenesis, then the answer to why life happened on Earth is it was because of chance. Since all the needed ingredients were in the primordial soup, a few strikes of lightning and a little boiling did the trick – so there was life. However, if we consider panspermia then it becomes intriguing especially if we are inclined to assume that advanced civilizations are the source of the contamination. Why did aliens do it? Why Earth? Why have not they come back to check on us?

Proponents of deliberate or directed panspermia forward the idea that advanced civilizations could have brought life on earth in order to colonize new planets and ensure the survival of the alien species. Francis Crick (the co-discoverer of

DNA) and Leslie Orgel espoused the concept of Earth undergoing directed panspermia by interplanetary or interstellar travelers.[1] When advanced civilizations have exhausted their resources or accidentally brought their planet to the verge of total destruction, it would be logical for these civilizations to look for and master new planets.

Humans have recently identified planets that are thought to be capable of maintaining life. However, we lack the technology to visit and test whether such planets are suitable refuge for our species or not in the event that Earth becomes inhabitable due to nuclear winter, excessive pollution or intense global warming. Advanced civilizations, on the other hand will likely have the means to transport themselves or to send probes to test potentially habitable planets. If they are highly advanced, they could even visit multiple planets to increase their chances of success. In this scenario, Earth will just be one of the many planets they might be monitoring as an experiment replication.

Another possible reason why aliens could have seeded Earth with life or its precursors is that they were just curious and that directed panspermia was just a part of their investigations to understand neighboring planets, galaxies and the universe. They could also be testing their theories regarding how life originated on their own planet. Therefore, Earth could be their sole *in situ* experiment or it could be one of the many replicates they are observing.

Some experts suggest that aliens want to study how humans are going to solve global warming to help them solve their own problem with this geoclimatic phenomenon. This could be likely but a little implausible if you consider that if they have started the experiment it means that global warming is already occurring on their planet and waiting decades or centuries for humans to solve climate change on Earth (or other beings on other planets) and THEN trying to apply these findings would be impractical or ill-advised, to put it mildly.

If advanced civilizations were the ones who started life on Earth as an experiment, then why have not they come back or visited us? There are several conceivable answers to this question. One is that they have already dropped by but it was too early for humans to witness it and they also did not leave any evidence of their presence. Another is that visiting is not in their plan and they would rather send probes instead.

Recently, there has been news about a conspicuous visitor to our solar system in the form of a comet - Oumuamua. The appearance of Oumuamua has been described as cigar-shaped and some researchers have raised the possibility of it being an alien spacecraft or probe.[2] However, evidence for such claims has been limited and Oumuamua has only left more questions instead of answers.

The other answer is that they could be simply observing us from afar (more on the zoo hypothesis later). They might have also lost their planet and obviously could not come back here. In this case, Earth can be therefore compared to an abandoned garden made by the Mayans before their civilization was wiped out by a disease or famine.

CHAPTER 2

Panspermia

> *"Where is everybody?"*
> – Enrico Fermi

Panspermia used to be the stuff of science-fiction for many biologists who view the hypothesis as either a great leap of faith or simply a convenient way of explaining life on Earth. Recently, however, more and more scientists have been welcoming the idea with an open mind. In fact, panspermia is no longer a distant possibility but is now being seriously discussed among experts as a probable origins scenario. In recent years, evidences and discoveries have been mounting to support the probability of finding alien life inside the solar system and even beyond.

Accepting panspermia means accepting that there is abundant life outside Earth. Life somewhere in the universe is thought to be not only probable but also likely. This is because there are just so many stars in the universe. More importantly, research by NASA has shown that at least half of the

1,000,000,000,000,000,000,000 stars in the universe harbor planets. That is 50% of 1 billion trillion of stars in the *observable universe*! Who knows how much of the universe that the world's most powerful telescopes have not seen or predicted through computations by astrophysicists? Planets beyond the observable universe could be teeming with life for all we know but because of the distance we might never hear from them given the possibility that even light could not reach us here while the human civilization exists. In fact, according to estimates from a study from the NASA Ames Research center, there could be tens of billions of Earth-like planets (habitable) in the Milky Way galaxy alone!

    Earth could be unique compared to the rest of the planets in the solar system but not that unique if you include all the probable planets in the universe in the equation. Surely, a bunch of other planets could have undergone the same processes as Earth, leading to the evolution of life. All the matter that make up the universe came from one enormous explosion called The Big Bang. This means that all planets that were formed through accretion when proto-stars formed into stars share common raw materials. The chemicals, therefore, that were present when life evolved on Earth could be found on other planets as well. The conditions that planets go through are also not that unique. In recent years, a number of planets, such as TOI 700d, have been identified to be lying around the "Goldilocks zone" (not too hot/close to the star but not too cold/far from the star), a location that is just right for the evolution and maintenance of life.[3]

While some might think that life on Earth arose from the occurrence of special chemical reactions, the elements making up this one-of-a-kind chemistry are fairly common. Carbon, hydrogen, oxygen and other elements comprising organic compounds are plentiful in the universe. In fact, amino acids, known as the building blocks of life, have been detected in a comet's atmosphere. Glycine, its precursor molecules and phosphorus were found in 67P/Churyumov-Gerasimenko's cloud of gas and dust.[4] These materials were believed to be enough to spark the evolution of life if exposed to a right amount of energy. Comets of this kind, thus, resemble inoculating needles (used by microbiologists) containing the prerequisites of life that are stored at cold, preserving temperatures. All that needed to happen was for comets to land on a soft, warm medium or substrate such as ocean water or a marshy land in order for the elements to be freed, to mix and to react with each other. It should be assumed, however, that the foreign ingredients survived the impact and intense heat generated by the comet passing through the atmosphere and its pancaking on the water surface.

Zoo Hypothesis

> *"People deny the presence of intelligent beings on the planets of the universe" because "(i) if such beings exist they would have visited Earth, and (ii) if such civilizations existed then they would have given us some sign of their existence."*
>
> – Konstantin Tsiolkovsky
> *The Planets are Occupied by Living Beings*, 1933

The zoo hypothesis has been forwarded by its proponents in order to deal with Fermi's Paradox - why we have not seen aliens when the universe, due to its old age and vastness, should be expected to host life in other planets as well. Why do we feel so alone in this universe? Proponents of the zoo hypothesis suggest that aliens can see us but they have made sure that we cannot see them. Extraterrestrials could have discovered the existence of Earth including all the life on it but chose to stay at a distance, not wanting to deal with us for some important reasons. One of the humbling reasons that have been suggested is that humans are not worth all the trouble for the aliens given the often destructive nature of our species as evidenced by the many wars we have waged and the ongoing human-caused destruction of our planet.

John Ball, the MIT astronomer who broached the idea behind the zoo hypothesis, believed that extraterrestrials maintained a hands-off policy with Earth in order to give humans the opportunity to pursue our own destiny.[5] This is quite noble for the aliens to do which is quite contrary to what some of us often fear: that aliens might be looking for us to obliterate our species in order to steal and colonize our planet and ensure their own survival. To explain this behavior of aliens, one needs to assume that they are way more advanced than the human civilization. They should also be quite self-sufficient and assured of their survival; thus they do not need the resources on Earth. Instead, these aliens are concerned with preserving the richness or diversity of life in the universe. This should sound like modern humans learning about the existence of a tribe that has not made any contact with the outside world. Or, like conservationists who are interested in studying chimpanzees in Africa but are doing so without the primates noticing them to prevent the animals from becoming too familiar with humans.

Therefore, humans and the other life forms on Earth could be living in a 'zoo' located in a corner of the universe far from other 'zoos.' Meanwhile, the sentient beings that are aware of the existence of these zoos are shielded from the residents of these holding places. These observers should be capable of interstellar travel and somehow made it impossible for us to see them because of the distance or by installing one-way barriers.

One might wonder, though, how long these sentient beings are going to keep us from learning about their existence. Should not they be curious about the experience and thoughts of humans? After all, humans have also developed philosophy and religion to try to make sense of our existence. If they have an illusion of grandeur making them believe that they are gods, then they would not be concerned about what humans believe in. But if they are just like us who have so many questions regarding why we are here and for what purpose we exist, then they would be inclined to talk to us. For example, Jane Goodall tried to live with chimpanzees in order to understand their behavior and gain insights about early hominids. No matter how advanced these aliens are they most likely evolved from earlier life forms just like us. They came from somewhere and they also need to discover their origins and having discussions with other sentient beings is certainly a start in the right direction.

Panspermia + Zoo Hypothesis

> "The universality of the genetic code follows naturally from an 'infective' theory of the origins of life."
>
> – Francis Crick and Leslie Orgel
> *Directed Panspermia*, 1973

Now, after introducing panspermia and zoo hypothesis, I would like to describe a scenario I will call the 'forgotten zoo.' If advanced, sentient civilizations were able to inoculate Earth with the starting materials of life or with primitive life itself, then the hundreds of billions of years from that initial contact up to now can be considered as quite a very long time. For us, it is definitely a long time because the modern human species, Homo sapiens, have only been on Earth for approximately 200,000 years (the earliest human species, *Homo habilis* emerged more than 2 million years ago).[6] On the other hand, the earliest forms of life have been reported to be around 3.5 billion years old. Based on more than one hundred years of modern science, there has not been any record of aliens – despite the fervent testimonies of "witnesses" – during the time since humans have existed and also during the prehistoric times. An alien space ship *might* be unearthed among the dinosaur fossils one day but right now, not a single evidence supports the idea that aliens have already visited our planet. If scientists were able to find the oldest fossils beneath the ground, then finding alien tracks, signs, fossils or spacecraft is also possible if not likely. However, alien enthusiasts lack even a single fossil to point to, which means that (1) there are no aliens to start with, (2) aliens have not visited Earth, (3) aliens already visited but no evidence remained, or (4) aliens deliberately removed any signs of their visit.

If we are to cling to panspermia and zoo hypothesis to explain life on Earth, then it is logical to ponder on the last two possibilities mentioned above. Then, we have to move on to a 'forgotten planet' scenario which implies that aliens seeded Earth with life or its basic components and then planned to go back to see if their experiment was successful but, over a large number of generations, forgot to come back to Earth. Alternatively, aliens could have remembered about Earth but, due to some reasons such as civilization decline, failed to travel through space to see what happened here. And after billions of Earth years, the alien species died out and became extinct. Considering these possibilities, it means that Earth life is either waiting to be discovered or is just 'on its own' in this part of our universe.

No one can prove any of these assumptions since there are no evidences to start with and we are talking about billions of years and space of unimaginable expanse that require tremendously advanced technology to surmount. Therefore, we are stuck in this place and could die out in the future without knowing if any of these possibilities is correct or not. Or, we could look at the evolution of life on Earth to see if any sign sticks out like a sore thumb, indicating that our ancestors were actually introduced on this planet and are thus serving a 'higher' purpose and that we are not just evolving 'aimlessly.' In other words, to get the answers we want, we need to first view life's evolution on Earth as an event that has a purpose. Evolution should appear as a deliberate development and not an accident in order to say that Earth is indeed a 'forgotten planet.'

So, we should be looking for signs in the evolution trend, which is challenging since evolution is an ongoing process and is, by no means, complete or will be completed in the future. However, there should be interesting facts about the long history of evolution. For example, were the odds of life springing up from primeval Earth too low? Was evolution too fast during some parts of Earth's history? Why there are so many viruses in the ocean, are they alive and what is their purpose? Did intelligence evolve quickly in humans? These questions could offer clues that can help support the suggestion that life on Earth was intentionally evolved although these could only remain in the end as speculation and nothing more. The best that we can achieve with this mental exercise is coming up with some kind of conspiracy theory. But, a conspiracy theory is better than nothing at all given the many centuries that humans have lived without knowing the purpose of our existence.

To reveal these clues, we need to review how life evolved during Earth's dynamic geologic stages. This entails briefly defining and illustrating life in the light of evolution, which could be helpful especially to those who are uninitiated. Knowing the characteristics of life can show how special life is. This book revolves around life so it is only fitting to understand its basic properties.

CHAPTER 5

Evolution of Early Life

> *"When ultra-violet light acts on a mixture of water, carbon dioxide, and ammonia, a vast variety of organic substances are made, including sugars and apparently some of the materials from which proteins are built up. [...] But before the origin of life they must have accumulated till the primitive oceans reached the consistency of hot dilute soup."*
>
> - J. B. S. Haldane
> *The Origin of Life*, 1929

What is life? It seems like a very mundane question but distinguishing living organisms from non-living things is not an easy task. Biologists have to grapple with a variety of definitions of life throughout natural science's long history. One of the characteristics commonly observed in living organisms is growth.[7] Growth is the observable manifestations of overall changes in the body of an organism - typically an increase in size. Through morphogenesis, organisms are able to attain the form that is specified by their genetic blueprint and allowed by their environment. Growth is obvious among plants and animals because we can see how eggs or seeds undergo morphogenesis to become looking like their parents. But, how about microscopic organisms? They undergo morphological changes, too, such as in the case of bacteria – albeit less remarkable compared to larger organisms. This kind of growth could be difficult for some to distinguish from the growth or formation of snowflakes or crystallization of salt although we know that these are not living things. Maybe, a way to settle this is to look at the source of this growth: living organisms grow by harnessing resources around them while nonliving things 'grow' when external physical and chemical conditions are just right. Snowflakes get their intricate patterns with the help of subzero temperatures and salt crystals attain their cubic form with the help of heat and dehydration.

Another trait of a living organism is the ability to replicate itself.[8] Reproduction or replication is achieved either sexually or asexually. We are familiar with how humans and other animals reproduce. We know about mating, pregnancy and birth of offspring. In the microscopic world, bacteria do not mate or combine male and female genetic material. Budding yeast simply undergoes, as its name suggests, budding to produce new daughter cells. It is not simply breaking off from the original body because there is mitotic cell division involved in the budding process. So here we can see that life intends to increase its number from one or few individuals. It is planning to perpetuate itself.

The next characteristic of living organisms is auto-repair or their ability to heal damaged or wounded tissues. From microorganisms such as *Stentor coeruleus*, *Xenopus oocytes* and *Chlamydomonas*[9] to plants and animals, unique auto-repair mechanisms can be observed. These mechanisms imply that their body has a way of knowing that an area or one of its parts has damage and needs to be resolved. Therefore, living organisms are somewhat self-aware and are able to heal wounds or injuries. Some might say that cars can now be designed with self-healing paint so they should be alive. We know that cars are not living things but self-healing paint and other similar technology are only inspired by nature and are surely interesting. Living organisms are also able to harness energy, a trait that enables them to not only survive but to also grow, develop and reproduce.[10] This means that life is an active player, not a passive one. We revert to the snowflake/salt crystal comparison. These nonliving things do not actively absorb and process nutrients and minerals for growth and reproduction. On the other hand, living things have special mechanisms for harnessing light (photosynthesis) and food (digestion). Not only are these special, these processes are also highly complex and have evolved through millions or billions of years.[11]

In other words, a thing must be able to grow, replicate, repair itself and harness energy in order to be considered as a life form. However, if one thinks seriously about it, life is just another representation of the chemicals around it. What makes up living organisms is present in the soil, air and water and are not in any way unique or novel. What is unique and novel about life is the seemingly perfect accumulation and organization of chemicals into organic compounds with specialized functions. It thus appears that life is a mirror of its surroundings. Life could be essentially an iteration of Earth itself.

How did this emergence of life happened on Earth? Scientists are convinced that organic compounds first formed then became self-replicating molecules and eventually led to the evolution of life billions of years ago. There are three hypotheses regarding the possible events leading to what we now know as Earth's biodiversity.

The first concept is called the prebiotic soup theory or primordial soup theory, postulated by J.B.S. Haldane in 1929. Prebiotic soup theory states that primordial oceans that existed 4.2 to 4.0 billion years ago (bya) provided the ingredients for the formation of the building blocks of proteins – amino acids. The chemicals from the atmosphere such as carbon dioxide combined with ammonia and water to give rise to sugars and organic compounds, mainly DNA and amino acids, with the help of the Sun's ultraviolet rays.[12] The organic compounds polymerized into complex molecules that were able to self-replicate. Over time, these self-replicating molecules were able to recruit and organize more chemicals around them, resulting in novel biochemistry and eventually evolving into living organisms.

The second concept is called the RNA world hypothesis proposed independently by three scientists: Francis Crick, Carl Woese and Leslie Orgel in the 1960s.[13] According to this hypothesis, RNA was the first nucleic acid that evolved as storage of genetic information and as catalysts of chemical reactions. Due to the instability and poor catalytic qualities compared to DNA and proteins, ribozymes became obsolete. Therefore, DNA and proteins became more widespread in primitive life forms.

The third concept is the "metabolism first" hypothesis proposed by Alexander Oparin in the 1920s.[14] According to this hypothesis, simple molecules such as acetate were spontaneously formed when carbon dioxide and water underwent initial reactions. Minerals and debris then catalyzed pathways for the synthesis of rudimentary organic molecules that later became the precursor of primitive forms of amino acids and lipids. These simple organic compounds then produced more of their kinds in ways resembling crude metabolism. Complex amino acids and nucleic acids were formed once metabolism was established, thus the "metabolism first" name of the hypothesis. These hypotheses all point to the likely evolution of the early forms of life. However, panspermia hijacks these processes by suggesting that RNA, DNA or proteins were delivered onto Earth from outer space. The primordial soup would still be useful in this scenario but the time required for evolving the earliest life forms should be reduced as a result. If organisms such as bacteria or their ancestors were able to hitchhike on asteroids or comets and landed straight on Earth, then the evolution time of life could be further reduced. It would be a bit unbelievable if organisms larger than bacteria or fungi were able to withstand asteroid or comet entry into Earth's atmosphere and its subsequent crash onto land or water. The impact could have torn apart tissues but bacterial and fungal spores could have fared better compared to multicellular organisms.

CHAPTER 6

The Importance of Water

> *"Water is the driving force of all nature."*
> – Leonardo da Vinci

When astrobiologists look for life outside Earth, they start with indicators of the presence of liquid water on the surface of planets, moons or asteroids. This is because we know that water was one of the crucial ingredients during the evolution of life on Earth. Without water, we would not have a primordial soup – unless there are water-like substances in other planets. But the odds of finding water in other Earth-like planets should be higher than finding other water-like substances. Why? Because hydrogen and oxygen – the two elements that make up water – are the first and third most common elements in the universe; between them (or 2nd) is helium.[15] If hydrogen and oxygen are abundant in the universe, then water should be abundant, too. This is what planetary scientist Li Zeng and his Harvard University colleagues found when they investigated the water content of some 4,000 exoplanets.[16] Water is VERY abundant in the universe so life could be a universal phenomenon, according to Zeng.

To see how important water is to life, we only need to consider our own body. The human body is 75% water by weight at the infant stage and 60% water at the adult stage. Most of this water is within cells and the remainder surrounds tissues or in the blood.[17] Except for some organisms such as tardigrades or water bears that are able to survive up to 30 years without water, most life forms will immediately die in the absence of water. Humans cannot last more than a week without water. This is because water is required in almost every biological process in all life forms. Enzymes catalyze virtually all the biochemical reactions inside cells and cannot function in their 3D structure without the help of a fluid such as water. Metabolic residues that need to be expelled from cells could accumulate and kill the cells if there is no water. Communication among cells and oxygen and nutrient transport in tissues will also be impossible without water.[18]

During the evolution of early life forms, warm water that lacked oxygen provided the medium for the crucial reactions. In addition, water's physical properties such as density, viscosity, diffusion and surface tension have been suggested to be helpful to primitive life forms as they are now to extant organisms. Water is important in facilitating protein folding, a necessary step wherein long amino acid chains attain their 3D structures. Proteins cannot function without such transformation.[19] The research team led by Gustavo Caetano-Anolles from the University of Illinois at Urbana-Champaign has revealed that between 1.5 billion years and 3.8 billion years ago, the function of proteins in archaea and multicellular organisms underwent gradual but drastic optimization. During this period, there was an explosive rise in protein domain reshuffling and architectures, thanks to an increase in protein folding speed. This proliferation of proteins, called the 'Protein Big Bang' by the research team, could have led to the emergence of eukaryotes on Earth.[20] Without water, none of these could have happened, according to the team.

Why is water important to a deliberate form of panspermia? Because from afar, advanced civilizations can detect the presence of water and assume that a planet could be hospitable to life. Earth, for instance, has a characteristically blue color when viewed from afar – thus the moniker, "the little blue planet." There are many little blue planets in the universe but not all of them have oceans of water like Earth. If we find a planet full of water that lies in the Goldilocks zone, then the possibility of it harboring life increases.

Advanced alien civilizations looking to colonize other planets would also look for the same kind of planet. A planet almost fully covered by water is a tantalizing prospect not just because life can potentially exist on it but because aliens can inoculate such a planet easily. They only need to send a spacecraft or a pod – samples safely deep inside it – to crash land on the planet. Given the wide swathes of water on it, there is a high probability that the object will land on water. The entry of the spacecraft or pod through the planet's atmosphere, if any, could burn up most of the object. The samples can be ejected automatically or damage to the space vehicle itself could let the seeds of life out to the water. A marshy land would be ideal but given the tendency of floating debris to wash ashore, the samples finding suitable substrates would not be that critical of a problem.

Earth's surface is 71% covered by water, information that might have delighted aliens looking to inoculate it. It does not matter that ocean water is salty. In fact, scientists have long suspected that life originated in the sea.[21] Some of the proponents of the life-from-ocean hypothesis point to the fact that all organisms use salt. However, this idea remains unproven and Earth's oceans billions of years ago were much saltier than oceans today. Ancient oceans were almost two times saltier than present seawater so these could not have been too conducive for life.[22] Nonetheless, there are a number of microorganisms that are considered halophilic or salt-loving, not to mention the countless species of marine life that have evolved the capability to tolerate or expel salt from their body.

Alien samples might also reach brackish to slightly less salty waters that could allow the survival of foreign life forms. Although we know that life on Earth today are sensitive to the effects of salt, it is possible also that the primitive, crude life forms might have been able to tolerate high salinity and evolved to give way to modern organisms that are stressed by too much salt in their systems. Nevertheless, if what the aliens sent to Earth were actually halophiles, then high salinity of the ancient oceans becomes a nonissue. Aliens could have known the properties of Earth's oceans and could have designed their inoculation mission accordingly.

Panspermia proponents need not look very far for the source of these halophilic ancestors of ours. It turns out, Mars has also had oceans with higher salinity than that of Earth's oceans. According to scientists, when Mars lost most of its water, the red planet's water became even saltier.[23] Now, what remains of Mars' waters is trapped within the planet's northern and southern pole ice caps. Between the equator and the north pole of Mars, subsurface frozen water is also suspected to exist. However, the presence of liquid water on our neighboring planet has taken many years to prove. What is clear, though, is that Mars once had liquid water and that its oceans could have hosted life – probably halophiles.[24]

An asteroid impact has been hypothesized to have caused the ejection of rocks from Mars' surface into space. These rocks are believed to have landed on Earth as meteorites known as nakhlites. Suffused with liquid water, nakhlites are viewed as possible vehicles of halophilic life forms that could have seeded Earth.[25]

The abovementioned scenario might be irrelevant to the forgotten planet hypothesis forwarded here. However, it illustrates the importance of water in the evolution of life on Earth. Having water, no matter how salty, covering most of our planet provides a number of possibilities for life to reach and proliferate here. Earth, in other words, is a blank canvas waiting to be painted with the vibrant colors of life. For microbiologists, Earth can be represented by a petri dish containing a nourishing medium waiting to be smeared with the early forms of life. And just like Alexander Fleming's forgotten petri dish, Earth could have been inoculated with alien life or its precursors but the experimenter forgot about it or simply failed to find out the outcome of the experiment.

CHAPTER 7

Evolution Leads to Biodiversity

"One general law, leading to the advancement of all organic beings, namely, multiply, vary, let the strongest live and the weakest die."
— Charles Darwin

The timeline of evolution of life on Earth started some 3.5 bya when the oldest form of unicellular organisms existed, although prokaryotes or cells without nuclei could have existed earlier. The next groups of organisms believed to have existed were unicellular organisms that were able to facilitate photosynthesis and those that were able to utilize methane. Viruses appeared around 3 bya although they could have arisen alongside the oldest unicellular organisms. Around 2.45 bya, oxygen became widely available in the atmosphere. Eukaryotes, or cells with nuclei, appeared more than 2 bya.[26]

Since then, various organisms appeared between 1.5 bya – 0.2 bya: protozoans, sponges, fungi and corals (in order). However, there was a rapid proliferation of new organisms during the Cambrian period (also called the "Cambrian Explosion") around 535 mya when jellyfish, echinoderms, myriapods, worms, crustaceans, mollusks, arachnids and arthropods came to existence. Bony and cartilaginous fish and other vertebrates appeared approximately 460 mya while tetrapods, other insects and amphibians first walked on Earth roughly 400 mya. Dinosaurs and other reptiles appeared 240 mya, followed by mammals around 210 mya and the first primates around 85 mya or earlier.[26]

The abovementioned narration of life's evolution on Earth gives an impression that the process was continuous. However, this was not really the case since there were several mass extinctions that occurred before we were able to witness the great biodiversity that we have now. In fact, it sounds incredible to be reminded that more than 99% of all life forms that ever walked (or grown) on the face of the Earth are currently extinct.[27] Some of these organisms slowly died out or gradually faded away after many millions of years but most of these extinct species suffered sudden extermination, thanks to mass extinctions caused by asteroid impact or supervolcano eruption.

The most famous of these mass extinctions is the Cretaceous-Tertiary extinction event (K-T event) which caused the end of the non-avian dinosaurs as well as other organisms approximately 66 million years ago (mya).[28] The K-T event is also viewed as the cause of the rapid diversification and evolution of mammals and birds that exist today. Before the K-T event, the global diversity of life on Earth was thought to be increasing given the abundance in nutrients as a result of climate change due to widespread volcanism. During this mass extinction event, around four-fifths of species on Earth suffered extinction. The dinosaurs are the most popular of these victims of the K-T event. Alongside dinosaurs were marine invertebrates that were proliferating during the Mesozoic Era but were eventually wiped out.

The most widely recognized theory explaining what happened during the K-T event is the "asteroid theory" or the Alvarez Impact Theory of Mass Extinction proposed by American scientists Walter Alvarez and Luis Alvarez. This theory claims that the impact of a bolide, which is either a meteorite or a comet, caused the K-T event. The impact spewed massive amounts of rock and dust into the Earth's atmosphere, covering the planet in darkness for several months. The debris in the atmosphere blocked sunlight, making photosynthesis impossible. Without photosynthesis the green plants died out, disrupting the entire food chain.[29]

Some animals such as lizards, snakes, turtles and crocodilians were also affected but were able to survive. The reason why most of those animals that survived the K-T event were small mammals is because they were able to avoid the heat from the asteroid impact by burrowing or swimming underwater. Also, the mammals' diet was mainly composed of aquatic plants and insects which were thought to be spared from severe depletion during the impact and subsequent lack of sunlight. On the other hand, dinosaurs were large, making it impossible for them to scamper away and avoid the heat from the bolide crash and their diet was mainly made up of terrestrial plants that were destroyed because of their inability to perform photosynthesis due to the protracted darkness.[30]

The K-T event was not really all in vain, though, in terms of biodiversity. This is because it has been observed by scientists that biodiversity typically increases following mass extinction events. There had been five mass extinctions on Earth and each event gave way to a more diversified biosphere. After the K-T event, for instance, a number of species of land plants, insects, birds, fish and mammals were found to have diversified. Experts attributed this to the ability of animals to innovate when offered different types of food sources and habitats that were left available by species wiped out by the extinction event. Also, new species emerged due to the opportunity given to survivors to explore different adaptations with the absence of competitors and predators.[31]

So, despite this interruption caused by mass extinctions – we might ask – has evolution of life on Earth been slow or fast? The answer is: it depends where we look. If we look around 3-1 bya, we could say that the evolution of life has been very sluggish. Why? It is because it took approximately 3.5 billion years for proto-cells to evolve into more complex forms like sponges. That is certainly slow if you compare how fast sponges evolved to give way to humans – 750 million years. This is just based on looking at the timeline of evolution recently discussed. Or, is it just an illusion?

One of the reasons behind the faster rate of evolution in recent time is the evolution of sex. With the help of sexual reproduction, every organism received a set of redundant genes, allowing an organism to survive even after a critical gene has suffered a mutation. Instead of dying because of the mutation, organisms can pass on genes containing modifications that could prove useful for the offspring's reproduction and survival. More importantly, recombination through sexual reproduction dramatically speeded up evolution.[32]

Despite the slow progress and the hiccups in the form of mass extinctions, evolution has been successful in introducing and maintaining biodiversity on Earth. However, one question we can ask by now is, "Is evolution predictable?" If we scorch Earth and wait for it to return to a certain state wherein life could evolve from scratch, would we see the same biodiversity that we are witnessing now? If we reset Earth to 4 bya, would we have similar sets of fossils as well as human civilizations as we do now? The answer appears to be yes, according to the research by Alex Pigot from the University College London's Department of Genetics, Evolution and Environment. Pigot says that evolution is predictable and that re-running the tape of life would result in the evolution of similar-looking organisms that are present today[33] (a chapter in Part Three is dedicated to predictability of evolution).

This finding regarding the predictability of evolution is certainly interesting, especially through the point-of-view of potential alien civilizations looking to start life on Earth or in other parts of the universe. They could have known or were testing this phenomenon by sending the building blocks of life to planets that can harbor life. They should know what to look for or what to expect when they seeded Earth. Just like any experiment in the laboratory, scientists have an idea what will happen but will usually test samples to confirm their hypotheses. For example, microbiologists can change genes in bacteria and let these grow on appropriate media in order to see that the genetic alterations match the microbes' eventual physical characteristics or phenotype.

Aliens could have been trying to understand their own evolution or how life started on their own planet. Upon finding that Earth is identical to their planet in many ways, aliens could have decided to recreate the emergence of their life form just like how Stanley Miller and Harold Urey experimented with water, methane, ammonia and hydrogen to simulate the conditions of the primordial soup.[34] But this time, aliens added some precursors patterned to their genetic makeup. They could have also added some accelerants or evolution catalysts along the way – probably during when the evolution is observed to speed up considerably. To entertain this slim possibility entails assuming that the alien civilization has time on their side. It means that the aliens can wait for billions of years in order to see the result of their experiment. This might be mind-boggling to some but let us make simple analogies to illustrate that this is not really entirely impossible.

Let us consider the case of a petri dish full of bacteria. Those bacteria could be us and we know that bacteria have a simple life cycle comprised by a lag phase, an exponential phase, stationary phase and death phase. For the bacteria this cycle could be a long period or a 'lifetime' but for aliens this is just a short period, maybe only hours or days for them. In the laboratory, microbiologists experiment with bacteria and usually leave them inside incubators overnight, checking the next day how the microorganisms have grown. Take also the case of experimenting with mice. Lab mice have lifespans that range from 1.3 to 3 years. For mice, three years is a long time but for humans who routinely test these animals, three years is not that long, obviously.

Thinking that there are different levels – with unimaginable orders of magnitude – to the perception of organisms of time might be difficult but it is only one of the aspects that need careful consideration in this exercise. How about intelligence? It is understandable for us humans to think of our species as the most intelligent form of life on Earth. But if there are life forms outside Earth, would we still be the smartest? If we can say that we have evolved quite ahead of fishes, birds, reptiles and other mammals then it is a big possibility that other alien species can say the same thing about themselves as well. What if these aliens were way smarter than us in a manner that we are way smarter than earthworms?

According to Charley Lineweaver from the Australian National University Planetary Science Institute, intelligence is not the priority of evolution.[35] Lineweaver has come to the conclusion that "evolution is about survival, not intelligence." But what if Lineweaver is wrong? What if intelligence is a trait that evolution eventually found to be very valuable and that other civilizations outside Earth also know about this? Also, if evolution is really that predictable, highly intelligent civilizations could have sowed planets around the universe with life to try to evolve or to test for similarly intelligent life.

PART TWO:
Reporter Genome

Evolution of Intelligence

"But nature is remarkably obstinate against purely logical operations; she likes not schoolmasters nor scholastic procedures. As though she took a particular satisfaction in mocking at our intelligence, she very often shows us the phantom of an apparently general law, represented by scattered fragments, which are entirely inconsistent. Logic asks for the union of these fragments; the resolute dogmatist, therefore, does not hesitate to go straight on to supply, by logical conclusions, the fragments he wants, and to flatter himself that he has mastered nature by his victorious intelligence."

– Wilhelm His, Sr.
On the Principles of Animal Morphology, 1888

Intelligence is often defined as "the ability to plan, reason, think abstractly, solve problems, understand ideas and learn." However, this definition only takes into consideration the human species. Other animals are intelligent, too, to some degree. Are plants, fungi, bacteria or viruses intelligent as well? We already know that humans have a very sophisticated level of thinking. Sometimes we wonder if this capacity is a fluke of nature because humans are conspicuously smarter than other animals. Cetaceans, elephants, pigs, octopuses and crows are quite intelligent but they do not stand any chance when compared to humans. Why the large gap?

Actually, the question regarding brain power disparity is not really that consequential if we think that evolution is a seemingly random process, if not for natural selection, towards more complex life. The more pressing question is "is life on Earth evolving to produce the most intelligent organism for some important purpose?" As just mentioned, this is not the case according to Lineweaver. However, it appears that intelligence in different species has been increasing over time, at least based on the vantage point of humans. Surely, looking at the past, humans are bound to think that we are the peak of the evolution of intelligence. Not a single animal species has achieved awe-inspiring feat of creative or innovative thinking – only humans have done that. If that statement was false, then we should now be looking instead at ancient civilizations that feature non-human animals. But nothing of such kind can be found among the fossils or ruins.

If we think that evolution has anything to do with humans having powerful cognitive ability, the past should offer clues how intelligence has been shaping through billions of years. According to some experts, the evolution of animal intelligence began with *tiktaalik*. *Tiktaalik roseae* is an extinct species of lobe-finned fish that lived during the Late Devonian Period and discovered in 2004. According to the book *Your Inner Fish*, authored by Neil

Shubin, *tiktaalik* is the missing link between fish and tetrapods. This species is said to demonstrate how vertebrates invaded land around 375 mya. By evolving limbs that propped up its flat head and propelled its body over land, the *tiktaalik* paved the way for tetrapod to roam new land territories thus pioneering the transition from water-dwelling fish to terrestrial animals.[36]

Scientists believe that this decision to conquer land is a sign of primitive animal intelligence. Some might argue that it could be a result of accidents wherein fishes get thrown out the water and onto the shore because of strong waves and these fishes just slowly developed the ability to wriggle and 'walk' on land over time. Nonetheless, persevering in such conditions and looking for new ways to find food and survive in new environment should qualify as an indication of a higher level of intelligence compared to that of its predecessor that were content with competing for food in the water with many other fishes. It is obvious that intelligence evolved in different kinds of animals because they provided advantages to these organism. Parrots, according to research, have been found to be capable of addition and subtraction. These birds can even speak in complete sentences.[37] Intelligence evolved in parrots because they have a need for it due to conditions in their habitat. Experts believe that because parrots are faced with the challenge of surviving in a forest being rapidly affected by human activities and environmental changes, they were forced to evolve intelligence. Their intelligence arose from the need to know where food will be in the coming days, thus requiring the parrots to predict the future.

So, can these other animals evolve to have intelligence equal to that of humans? If given enough time it might be possible for birds to have highly intelligent species of their own. What is clear, though, is that humans were able to reach extreme heights in intelligence because everything went well for us. These include appropriate diet, blood system and dexterity. All groups of animals have the capacity to evolve intelligence but in humans, the process was relatively quicker. Hominids are omnivorous and their diet was able to feed an ever-growing brain.[38] Hominids can breathe air which made it possible to support large brains. Hominids also have dexterous hands that enabled them to fashion tools and perform various activities.[39] Furthermore, hominids lived on land and were able to harness different kinds of resources such as wood and stone. Later, humans were able to achieve more advanced technological feats because bronze, iron, ceramics, glass, carbon fiber, silicon and rare earth minerals are available to humans or could be produced on land. Obviously, the same cannot be said if cetaceans wanted to build advanced technology underwater. Maybe they might be able to build submarine domes so we really never know!

As expected, intelligence has been discovered to be controlled by a set of genes in the human genome. Danielle Posthuma and her colleagues from Vrije Universiteit Amsterdam, Netherlands, have found 1,016 genes associated with human intelligence.[40] Interestingly, her team also reported that the genes linked to intelligence are likewise implicated in maintaining cognitive health. Smartness genes were negatively correlated with depression, attention deficit/ hyperactivity disorder

(ADHD), Alzheimer's disease and schizophrenia. What is more intriguing, though, is that intelligence genes were positively correlated with autism and longevity. It is intriguing because extreme intelligence of child prodigies has been observed in autistic individuals, particularly those with Asperger's syndrome.

If we are to dare assume that if the right genes for intelligence are all received by an individual and that this individual is also gifted with longevity, then this individual could become a fine specimen for extreme intelligence and productivity. In other words, a community or a country of this kind of super-intelligent humans can create many discoveries and engineering achievements for the benefit of their kind. But what would these gifted people build?

Throughout history, humans have demonstrated the capability to form civilizations and to make useful technologies that include roads, bridges, cars, ships, high-speed trains and spacecraft. At present, we are pushing the boundaries of space travel. We have sent spacecraft (Voyager 1 and Voyager 2) that can propel itself away from Earth and past the planets of the solar system.[41] There are also plans to colonize Mars and public-private partnerships have started designing and testing rockets that will bring people and building materials to the red planet. Powerful telescopes have been scouring the universe for Earth-like planets and any sign of life.

But, isn't it perplexing that we are hell-bent on exploring deep space when we have not fully seen what lies below Earth's oceans as if we are so sure that what we are looking for is not miles below sea level? What are we searching for anyway? Why are humans so obsessed with what lies out there? Many would say that we humans are looking for answers to questions regarding our existence. But, why do we want to know about our origin? Why can't we find contentment here on Earth? Space exploration is technologically challenging, not to mention, very expensive. Yet, we pursue scientific discoveries when more pressing issues on Earth abound. Why can't humans just sit and relax and experience all the pleasures that life can offer? Why do we risk our lives and limbs in search of the unknown? We might as well answer why humans are very curious in the first place. Is curiosity in our genes?

Curiosity Gene

"The important thing is not to stop questioning. Curiosity has its own reason for existence. One cannot help but be in awe when he contemplates the mysteries of eternity, of life, of the marvelous structure of reality. It is enough if one tries merely to comprehend a little of this mystery each day."

- Albert Einstein

There is, indeed, a "curiosity gene." That gene is called DRD4, which stands for dopamine receptor D4. It was identified by Max Planck Institute researchers in 2007. DRD4 is responsible for building receptors for the neurotransmitter, dopamine, in the brain.[12] The existence of a gene for curiosity indicates that this behavior is an adaptive trait. It is interesting to note, however, that DRD4 was not discovered first in humans but in a passerine bird, *Parus major*. This gene was observed to be involved in a strong exploratory behavior in the songbird, the great tit.

In humans, a variant of DRD4 was found to be associated with a risk-taking phenotype. Extraversion and novelty seeking behavior among skiers and snowboarders was reported to be linked to the -521 C/T variant of the curiosity gene.[43] The authors noted, though, that this gene variant was not associated with impulsive sensation seeking.

One of the most relevant books discussing the role of curiosity in human evolution is Alexandros Kourt's *The Curiosity Gene*.[44] According to this book, curiosity is responsible for the survival of humans as well as for our evolutionary advancement to become the most intelligent creature on Earth. Kourt claims that curiosity is the missing link in the story of evolution of humans. Intense curiosity is a unique characteristic of humans, setting us apart from other animals. The origin of human curiosity can be traced back to 2.3 mya, finding its way into the human genome and driving our success as a species. This book is an interesting look at the importance of curiosity in human's evolutionary history. However, Kourt stopped short at the suggestion that curiosity holds the key to our happiness and personal fulfillment – that's it. Curiosity is a means to an end which is survival. But, what happens if you stretch the usefulness of curiosity beyond such primal application? Can it point us to an answer regarding the question, "where is evolution headed?"

Curiosity appears to have a more important purpose if we view evolution in the eyes of alien civilizations that are studying how organisms emerge on Earth. In this scenario, curiosity is a tool to help organisms to rise above everything else to look for their real origin or their 'creator.' Humans are not just content with satisfying their curiosity, curiosity drives humans incessantly towards wherever it leads us.

Discovering what lies beyond Earth might not be beneficial to us in the short term but why do we still do it? If other members of another intelligent species of animals see us investing invaluable resources to explore outer space, they might think that we are crazy because no apparent benefit will come out of such an exercise. They might even laugh – if they are able to laugh – at NASA scientists for naming their Mars exploration rover *Curiosity Rover*! They would likely ask, "Can curiosity feed you or make your species better at reproduction?" Of course, we would answer, "No, but its results might – in the future."

According to Richard W. Byrne from the Centre for Social Learning & Cognitive Evolution and Scottish Primate Research Group, information gathering, which is basically what animals do when we are satisfying curiosity, is worth the effort even if there is no clear reward for doing it - *unless* it is unreasonably costly or risky.[45] For humans, this does not appear to make sense because, as we know, (aside from the fact that building rockets is literally very expensive!) venturing into space that lacks oxygen and gravity – not to mention teeming with radiation – is certainly costly and risky.

Evolutionary biologists suggest that human curiosity is associated with our trait called neoteny which means 'retention of juvenile characteristics.' According to this concept, the human species is more child-like compared to other mammals and this is demonstrated by our being relatively hairless and having a big brain relative to our body size. Therefore - proponents of neoteny claim – the playfulness and curiosity that humans display throughout our lifetimes is a behavioral manifestation of neoteny. Also, neoteny is considered as an evolutionary shortcut resulting in humans being more juvenile than non-human primates, making us weaker than our relatives. However, it also made us highly curious organisms with great capacity for learning.[46]

If we accept neoteny as the reason behind the extreme curiousness of humans, then we can say that humans are just literally wandering about on this planet with no apparent direction or purpose. And we can stop there because human evolution has reached an obvious end to its story. But we can also say that life has been designed that way – to continuously search for something interesting, whether on this planet or somewhere else.

However, if we consider humans as having more special or advanced cognitive abilities compared to other animals such that we can override basic evolutionary laws, then curiosity takes a different context. It would then seem that curiosity is a proof that humans are looking for something else, something that could explain or complete our existence. It is as if we have another life cycle and that part of the cycle is not yet found or completed by humans on Earth (for now). So we keep on searching for that missing piece. How else can you explain the purpose of extreme curiosity in the big evolutionary picture?

This human's search for an apparent next stage reminds us of the story of zombie snails. Zombie snails are called so because they behave like mindless snails, attracting predators instead of hiding from them. This behavior is due to an infection by *Leucochloridium paradoxum* or the green-banded broodsac. *L. paradoxum* is a parasitic flatworm that typically infects gastropods, like the land snail *Succinea putris*, as an intermediate host.[47] The parasite starts as a larva inside the host and grows into a sporocyst in the snail's digestive system. Sporocysts then develop into broodsacs containing numerous cercariae. These broodsacs travel and invade the snail's eye stalk. This turns the eyestalk into an eye-catching structure – especially for predators – due to its colorful, pulsating and swollen appearance resembling that of a wriggling grub or caterpillar. The broodsacs' invasion of the eyestalk causes the eye to lose its ability to perceive light intensity so the snail tends to go to open or elevated areas where there is more sunlight since the snail most likely finds all the other places much darker than they really are due to the broodsacs obscuring its eyes. Staying in open or elevated areas exposes the snail to higher chances of being eaten by birds. Inside the digestive system of the bird, cercariae become adult diastomes that are capable of reproduction. Diastomes then lay eggs that are excreted by the bird along with its droppings. The cycle is then repeated when snails consume the droppings containing the *L. paradoxum* eggs.

    This analogy does not imply that humans are zombies that are fully unaware of what is going on around them. This is just to illustrate that there could be a deeper process that could be operating inside us. Obviously, we appear to be healthy and our minds are not clouded by any overt foreign body like the broodsacs but is nonetheless causing us to behave in weird ways (such as wandering aimlessly in the name of curiosity). But what if what is inside us do not appear as foreign? We do not appear to have strange organs or body parts that can control our decisions or actions. What we have, though, is bits of foreign DNA in our very own genome.

Bacterial genes have been found to be introduced into human somatic cells, cells that do not form reproductive cells through lateral gene transfer. David Riley and Karsten Sieber published the result of a study indicating that more than 7,000 events of lateral gene transfer have occurred from bacteria to humans. Interestingly, though, almost all of these lateral gene transfers were found to occur in cancerous cells.[48] A similar mechanism has long been known to exist with viruses and insertion of viral DNA into human somatic cells has been directly implicated with cancer development.

Unlike bacteria, viruses were able to insert their genes into the genomes of human egg and sperm cells. Frank Ryan's book *Virolution* extensively discusses this phenomenon.[49] The virus DNA can be passed down from the parents to offspring, allowing the integrated virus DNA to be replicated and reinserted into the ancient human genome. After thousands of generations and hundreds of thousands of years, 8% of human genome is now made of viral genes. This is quite interesting given that the original human genes that serve as the blueprint for our body's overall development and functioning make up a meager 1.2% of the human genome! Do viruses hold a special place in the evolution of life on Earth? For example, did viruses have a role in the evolution of the most sophisticated organisms on this planet, us humans?

CHAPTER 3
Evolving the Candidate Species

"The cosmos is within us. We are made of star-stuff. We are a way for the universe to know itself."
— Carl Sagan

If we are to stand in the shoes – if they have any – of the alien scientists the purpose of their experiment can become clearer. Why should anyone make an evolution experiment on a distant planet? Aside from curiosity, which is a strong driver in itself, we could also be motivated by all the genetic combinations that life on Earth could come up with. If we share the same genetic framework, then such novel genes on Earth could be considered a huge treasure trove for us (the alien civilization) for advancing our own health, medicine and related scientific research. We could be searching for the fountain of youth or looking to propagate species we already lost on our own planet. We also could be waiting for an advanced civilization on Earth to develop unique technologies that we do not have, although this is quite improbable. Highly advanced civilizations capable of zipping around the far reaches of the universe and colonizing other planets should have been able to develop all the technologies that they need.

    Nevertheless, alien scientists could learn a lot from any new civilization that they study. Since they should know that evolution is predictable and that they could have accelerated the process, all they have to do is wait for an intelligent species to rise up from all the organisms on Earth. Again, we are assuming here that the experimenters' civilization is stable and they have extremely long life-spans. As mentioned earlier, a good analogy of the lifespans would be between a microbiologist and his/her bacteria or *Caenorhabditis elegans* (a worm) samples: the microbiologist as the alien scientist and the bacteria or *C. elegans* as earthlings. This alien species would have to develop the technology to make deep space travel possible to perform such an experiment.

    If the alien scientists were able to start multiple experiments similar to what they have done on Earth, then they can theoretically just expect for any of the candidate species - like a highly intelligent and highly curious civilization such as humans - to reach their planet to confirm that a particular experiment was successful. This could also be the reason why they did not bother to return to Earth and assess how the experiment went. In other words, a failed experiment is not worth another trip to our planet and the failure is quite clear from the nonarrival of a reporter species. The alien scientists only need to have a positive result and that result is a self-reporting specimen to complete their experiment. In the case of planet Earth, that reporter species could be none other than us humans.

We might not know it but we could be on our way to letting the alien scientists know of our existence. Just like the zombie snail discussed previously, humans are wandering around our galactic neighborhood and are thus bound to attract attention. We are exploring our nearby planets and moons, sending probes beyond the solar system and broadcasting signals all around, hoping that advanced civilizations are listening or looking for us.

If this forgotten intergalactic experiment turns out to be true, then it is clear that we are the candidate/reporter species for this planet and we are performing our duty quite well. Since our technological prowess continues to improve in leaps and bounds, we are destined to eventually have the capability for intergalactic travel to finally meet the alien scientists.

Of course, after humans have left Earth or died out, other species could take over and continue the search for extraterrestrial life. That is, if humans failed to do the job. The next intelligent species could come from terrestrial animals, although in what animal species it would be difficult to predict. It could take a long time for another species to reach what humans have already achieved but as long as the alien civilization exists, the experiment continues. The alien scientists can rest assured knowing that evolution is predictable and that life tends to evolve towards intelligent forms.

When humans explore the universe, we would be bringing with us all the advanced hardware and software that run the complex machines and spacecraft that enable our interstellar travel. Aside from these, we would also be carrying with us a wide variety of bacteria, fungi and other microorganisms on our belongings and inside our bodies. Expeditions looking to colonize other planets would also likely have a stash of seeds and live or cryopreserved gametes of livestock for establishing agriculture on prospective home planets. Explorers might also bring encyclopedias or databases about Earth and everything on it. In short, we would be transporting a lot of information about our home planet and this is just what the alien scientists need. Based on the clues we might give the aliens, they could determine our species' origin, as well as the exact location and the current status of our planet.

If they are only after new knowledge or satisfying their curiosity, then aliens knowing about Earth and its resources would not be that catastrophic for humans. However, if the ultimate objective of the alien civilization is to eventually exploit new resources that humans heavily depend on, then it could get complicated for us. Either way, humans making contact with another civilization outside the solar system could invariably lead to our species and our planet being unnecessarily exposed to outsiders. In other words, our space exploration activities could end up in a bioprospecting expedition to Earth by other alien civilizations.

Even if the human explorers are careful enough not to include books and databases in the cargo while looking for alien life, they could still divulge a lot just by presenting themselves to the alien civilization. This is because we are a walking encyclopedia, although we are not talking about our brain contents here. I am referring to a database billions of years old! This database is contained in an organic body capable of replicating itself. We humans are a walking, talking storage of vast information about planet Earth. We are, in a way, Earth's primary record keepers, as what the following chapters illustrate.

CHAPTER 4

Human Biology Reflects Habitat

> *"It is not the organs—that is, the character and form of the animal's bodily parts—that have given rise to its habits and particular structures. It is the habits and manner of life and the conditions in which its ancestors lived that have in the course of time fashioned its bodily form, its organs and qualities."*
>
> — Jean-Baptiste Lamarck

If the alien scientists are to examine the human body (with or without our permission), they are going to discover a wealth of information about our home planet. Our lungs are a dead giveaway about the presence of an atmosphere on Earth. The alien scientists can determine further the content of this atmosphere by looking at our blood. The bright red color of human blood results from oxygenation. This distinct color of blood develops when hemoglobin – a protein in red blood cells – binds oxygen.

Our digestive tract is also very revealing about the human diet, as well as the vegetation on Earth. Alien scientists can imitate how anthropologists surmise the behavior of our human ancestors. Our flat but thick-enameled molars give a hint that these were used for grinding seeds as well as for crushing marrow out of bones. The presence of different enzymes, such as amylase, in our saliva indicates the presence of starch in the food we eat. Our bone structure and even our heart's pumping capacity can indirectly indicate the strength of gravitational force on Earth. Research has found that the human tibia can handle up to 90 times the gravity of Earth. Meanwhile, the human heart has been said to be able to tolerate up to 5_{gE} or five times the gravity at the Earth's surface (about 4_{gE} in the long run).[50]

Of course, it is easy for the alien scientists to deduce that we thrive on land and not in water since we do not have gills, webbed hands or webbed feet. They would easily notice that we are bipedal; that is, if they found the human explorers alive and not dead or cryo-preserved in deep space. However, these are simple pieces of information that the aliens might gloss over because these are facts applicable only to humans and our environment. What they would probably be looking for is a bit of comprehensive but detailed look at our planet, including its history.

For example, if they are lucky enough, they could find a pregnant crew member of the human space exploration team. Using advanced equipment, the alien scientists would be able to observe the development of the human embryo inside the mother's womb. From the different stages of human embryogenesis, the alien scientists could somehow guess how the relatives, such as other vertebrates, of humans develop. They would notice that the human embryo temporarily develops tails but only retains the coccyx unlike all other vertebrates that develop fully formed tails. The human embryo also shows gill slits that are shared with other vertebrate embryos. The gill slits of human embryos end up forming the lower jaw and Eustachian tube, however, and do not completely become gill openings such as those in fish and larval amphibians.

From these observations, alien scientists would be able to deduce the existence of distant or primitive relatives of humans having tails or gills that are useful for dangling from tree branches and breathing underwater, respectively. However, this type of analysis would require a lot of guesswork on the part of the alien scientists who presumably do not have much information to start with. Therefore, they would have to use more reliable data to get a glimpse of the evolutionary history of humans. For that, the alien scientists would have to take a closer look at our genetic material.

Since they are an advanced civilization, they would probably know about DNA, RNA or any kind of genetic material familiar to them. We can assume that alien scientists are going to zero in on genes or genomes to know more about humans as well as Earth and all the other organisms we are related to. This is not entirely surprising since they could have done it on their own species and every other organisms on their own planet. They would most likely have their own tree of life and one of the branches could even lead to us! That last one is a big leap of faith but it would make sense if they indeed experimented on Earth and seeded life here. However, there is also a possibility that life on Earth could be totally different from that on their planet due to the unique geological conditions on either planet. This is inconsequential, however, and the only important part is that the alien scientists would know how to analyze our anatomy including our genetic characteristics.

CHAPTER 5

Genes Mirror the Environment

"Fossil bones and footsteps and ruined homes are the solid facts of history, but the surest hints, the most enduring signs, lie in those miniscule genes. For a moment we protect them with our lives, then like relay runners with a baton, we pass them on to be carried by our descendants. There is a poetry in genetics which is more difficult to discern in broken bones, and genes are the only unbroken living thread that weaves back and forth through all those boneyards."

- Jonathan Kingdon
The Self-Made Man: Human Evolution from Eden to Extinction, 1996

A reductionist view of the gene would represent it as an entity that enables the synthesis of a protein with a specific function which could be either structural or biochemical. A structural protein is involved in the construction of the physical aspect of cells or organisms. On the other hand, enzymes are proteins that catalyze chemical reactions in cells. In other words, genes can shape the structure and function of cells and life forms by encoding proteins for either structural or biochemical purposes. However, genes cannot determine the organism's physical attributes on their own.[51]

The environment of the cell or organism is a critical factor in the gene's performance of its role. It is also the environment that provides the raw materials for the formation of an organism's components. For example, in order for mollusks to make their own protective shells, they need the appropriate genes as blueprint as well as raw materials for the construction of this biological structure. The mussel obtains calcium ions from the environment by way of their gills, epithelium and gut. Calcium ions are transported to the calcifying epithelium with the help of the mollusk's hemolymph. There, the crystallization starts in order to eventually form the mollusk shells' mainly calcium carbonate structure.[52]

The abovementioned formation of the mollusk shell is a fine example of how genes depend on the environment to achieve their purpose. It also illustrates the fact that genes exist in tandem with environmental inputs. Genes can also be viewed as a means for organisms to build orderly structures out of disorderly raw materials from the environment. In a sense, therefore, genes somehow serve as a reflection of the environment where an organism thrives. But where did genes really came from?

Various hypotheses have been floated since the early 20th century to explain the origin of genes. One hypothesis suggested that during the duplication of DNA by cells, some genes were accidentally copied twice. We may recall that RNA or DNA could have emerged from the primordial soup to start the evolution of life on Earth. Thus, cells had the material for the storage of information as 'genes.'

The aforementioned hypothesis, however, does not explain how the very first gene came to be. Nonetheless, gene duplication has been confirmed by later studies to be indeed behind the proliferation of most of the genes we find today.[53]

These duplicated genes gradually formed gene families over millions of years of evolution. However, the respective function of many of these genes is not set in stone. Directly after duplication, gene pairs typically have the same sequence and function but through long periods of evolution, some of these genes undergo changes in their function. One example of this is the gene for hemoglobin. Hemoglobin is responsible for storing oxygen in red blood cells as well as for delivering this oxygen to cells and tissues throughout the body. According to studies, hemoglobin is a member of a family of genes that perform various functions involving oxygen. Hemoglobin, it turned out, evolved from proteins that sequester free oxygen molecules within cells to prevent them from causing damage.[54] Examples of genes coding for hemoglobin include *HBA* and *HBB*.

Scientists used to believe that gene duplication was the mechanism behind the existence of all genes. All these genes today are thought to be the product of countless series of duplications of a set of the original genes possessed by primitive organisms. This means that if we compare a gene sequence to all the other gene sequences that exist, we are bound to find a match or a distant relative of that gene. However, it appears that this is not always the case. Scientists have also found solitary genes – genes that exist in only one species. They call these genes 'orphan genes.' Further research has shown that orphan genes appeared more recently than the primordial genes and that orphan genes were not purely a product of billions of years of gene duplications.[55]

It turns out that there is more than one way that orphan genes can be generated. One is through gene duplication and subsequent rapid evolution. Rapid divergence of duplicated genes blurs the similarity between their sequences such that an orphan gene can be created. Another way is through the formation of new genes from noncoding DNA, indicating an evolution from ancestral genes without any function. These new genes, as the name implies, are called *de novo* genes.[56] Still another way of generating orphan genes is through horizontal gene transfer, a mechanism already described above.

Despite these discoveries, the fact remains that genes – no matter how old or how new – are also a function of the environment. Without any input from the surrounding physical or chemical entities, genes cannot perform its work. Therefore, genes are a reliable indication of the existence or involvement of an interaction between an environmental factor and a cell or organism. This means that we can use a particular gene to deduce the environmental conditions wherein the gene plays a role. For an example, we go back to the lowly but revealing mollusks. The pearl oyster *Pinctada fucata*[57] and the pearl mussel *Hypriopsis cumingii*,[58] share a gene called *Pif* which codes for a matrix protein known for regulating nacre formation. Nacre is a term used for mother-of-pearl, a biomineral found within the shells of mollusks. Hexagonal platelets of aragonite make up nacre. Aragonite is a form of calcium carbonate and is produced through biomineralization with the help of *Pif* gene. Studies have shown that this gene is strongly expressed in the pearl secreting tissues; in fact, *Pif* has been identified as a suitable gene for use in pearl farming.

Calcium carbonate makes up 4% of Earth's crust and is a ubiquitous material found in limestone, chalk and marble. As already mentioned, it is also a major ingredient of the shells of mollusks, including aragonite of mother-of-pearls. Knowing that seashells and mother-of-pearls are formed through the expression of *Pif* gene in mollusks helps us to connect the gene to the ocean where these organisms live. Calcium ions, required for calcium carbonate formation by mollusks, are a major ingredient of seawater. We can therefore associate the presence of *Pif* genes with ocean-dwelling organisms. However, there are freshwater-dwelling snails that can also make shells made of calcium carbonate. Land snails such as *Limax maximus* can also generate calcium carbonate-rich shells. *L. maximus* obtains calcium through its diet of wood, leaves, fungi, algae and animal carcasses, to name a few. Land snails also regularly consume soil that contains calcium. It has been reported that the snail is also capable of absorbing calcium by way of the sole of its foot. It turns out, gastropods similarly have a *Pif* gene which, along with those of the bivalves,' could have been derived from a gene that was present in a common ancestor.[59]

Given the abovementioned observations, when alien scientists are presented with a gene like *Pif*, they could hazard a guess that the organism likely possesses a calcium carbonate-rich shell and that it thrives in a habitat where calcium is readily available. This organism could either be found on land, in a fresh body of water or in the sea, but most likely in the ocean.

We might ask, 'how would the alien know that a gene codes for a *Pif* protein?' Well, calcium is not a very rare element in the universe. In fact, it is abundant in our own galaxy, the Milky Way.[60] There are even supernovae called 'calcium-rich supernovae' because they have plenty of – you've guessed it – calcium.[61] For instance, the supernova remnant Cassiopeia A which is 11,000 lightyears from Earth, contains a lot of calcium – approximately 2.0×10^{28} kg of it.[62] Thus, it stands to reason that wherever in the universe that life happened to evolve, there would likely be living entities that exploit calcium. And if life on the alien scientists' planet shares a genetic code with life on Earth, then it is probable that a gene similar to *Pif* exists in the alien's databases. A positive gene sequence comparison match would therefore tell the alien scientists: *Pif* = with calcium carbonate shell = most likely an ocean-dweller. Then, they could use other available genes and their environmental functions to extrapolate, confirm or debunk their hypothesis.

Compendium of Earth and Life Histories

"If you take the entire living biosphere, that's the assemblage of 20 million species or so that constitute all the living creatures on the planet, and you have a genome for every species the total is still about one petabyte, that's a million gigabytes... And somehow mother nature manages to create this incredible biosphere, to create this incredibly rich environment of animals and plants with this amazingly small amount of data."

- Freeman Dyson

If genes are a reflection of the environment of a cell or organism, then what do we make of an entire genome? An entire record of the environments where the cell or organism moves around and interacts with - that is what it is! Let us review: *Pif* gene indicates the availability of calcium ions in the marine environment of bivalves. Hemoglobin genes *HBA* or *HBB* signify the presence of oxygen in the air that the gene-bearer breathes. The gene for amylase, *AMY1*, hints the presence of starchy plants in the vicinity of the organism. These are just three genes but if they are found in one organism, they would reveal a lot about that organism's probable habitat. Imagine the different genes present in each bacterium, alga or fungus. The genomes of these organisms could provide snapshots of their environment. If one gene represents one photograph, then a single organism would generate an entire mosaic describing the organism's habitat. Now, if we post on a continuous, circular wall all the mosaics that each earthling has produced, placing those from the same type of habitat or location together with the help of an algorithm, then we can create a rough representation of Earth using all of the genes that ever existed.

Of course, not all of the genes in an organism code for the physical or chemical attributes of the environment where it lives. There are genes that are involved in various other internal processes like reproduction and homeostasis. So, let us take a look at the organism with the smallest genome ever known – the bacterium *Carsonella ruddii*.[63] *C. ruddii*'s genome is so small, it is only made up of only 160,000 base pairs. Compare that to the human genome which has 3 billion base pairs. This small size of the *C. ruddii*'s genome has led some scientists to say that it is not enough for survival since according to estimates, 400,000 base pairs is the minimum genome size for survival. But, guess what – this bacteria is alive. The interesting part is *that C. ruddii* has managed to reduce its genome to the bare essentials by designing around 90% of its genes to overlap with each other. Talk about efficiency! The downside, however, is that *C. ruddii* cannot survive on its own. Just like viruses, *C. ruddii* needs a host to provide it energy for its survival. Nonetheless, half of its genome is allocated for basic cellular processes such as handling and building amino acids into proteins – a function that is lacking among viruses. So what happened to *C. ruddii*? According to research, bacteria like *C. ruddii* can afford to shuttle its own genetic material over to its host given that it can survive just by feeding off of its host. Thus, these bacteria could have considered genes required for autonomous living as excess genetic baggage and thus expendable. In the case of *C. ruddii*, millions years of evolution and gene transfer has led to an organism that gave away many of its genes and is therefore left with a very small genome that is unable to support survival outside a host.

The abovementioned findings mean that we can have 80,000 base pairs from the most basic bacteria and still be able to build genes for detecting and exploiting its environment. For bacteria, exploring the environment entails having sensors instead of eyes, ears, nose, hands and feet that we and other animals have. So how do bacteria "sense" their environment? Bacteria use two-component systems to detect light and chemical signals around them. The system is very basic or crude.[64] First, a bacterium employs a sensor kinase to detect signals from the outside world. Once a signal is detected, a process called phosphorylation is initiated, activating a response regulator that causes a gene to turn on or off, and *viola*! – the bacterium can now see its world.

Now, imagine more complex organisms: algae, protists, fungi, plants, worms, insects and higher animals. Their sensors have certainly become highly sophisticated through billions of years of evolution. If bacterial organic cameras started from being able to detect dark and bright, along the way the cameras of other species evolved into black and white analogue units until humans came along with high definition, Technicolor, digital cameras. That is just for the eyes. Animal ears evolved from very primitive airborne sound-detecting organs of our terrestrial vertebrate ancestors that lived 100 mya during the Triassic period.[65] So do the highly discriminating noses of dogs, boasting 300 million olfactory receptors that could be traced back to our common ancestors with fish-like chordates called lancelets that existed approximately 700 mya.[66] Our skin's sensory cells and nervous system that we use to determine the presence, texture and temperature of things around us have its origin in the primitive nervous systems of Cnidarians such as corals, jellyfish, sea anemones and other aquatic animals.[67] In other words, the current set of genes that enable the functioning of our different sensory organs have a rich evolutionary history spanning millions to billions of years.

The genes that make our eyes, ears, nose and skin work have been changing through the years, all the while being improved, overhauled and polished by the different organisms that had or still have them. Through sequence analysis, we can determine which genes are the oldest and which ones are newer. We can also pinpoint the last common ancestor that had the earliest form of these genes. A phylogenetic tree or an evolutionary tree can show the evolutionary relationships among different organisms based on these genes.

The evolutionary history of life is also the history of Earth itself because of the association between genes and the environments as just discussed. Therefore, if an advanced extraterrestrial observer gets ahold of an evolutionary tree, that being could likely generate a picture or an idea about how Earth looked like in the past or how it has been changing as time passes by. So, the alien scientist only needs the information about our genetic material and everything about genetic materials to deduce what Earth looked like or looks like, from afar (millions or billions of lightyears away).

The most fascinating part is, even without a phylogenetic tree at hand, an alien scientist can still discover facts about Earth based on one highly-evolved, modern extant species – say for example, humans. The human genome can tell a lot not just about our species and our relationship with other organisms on Earth. Our genome can also serve as a standalone historical record of our planet. How did this happen? The clues are in the different genetic or evolutionary processes already described in the previous pages.

One of these clues is horizontal gene transfer. As has been mentioned, bacteria and viruses regularly facilitate HGT between them and with their respective hosts. We may recall that HGT has been implicated in the production of orphan genes. HGT has also been suggested as one of the probable mechanisms behind the acceleration of evolution rates among early prokaryotes (discussed in Part Three, Chapter 1).

Now, it appears that HGT is a significant contributor not only to the evolution of microorganisms but also of many animal species. In humans alone, 128 genes from foreign species' genes have been identified. Many of these introduced genes were involved in metabolism and originated from bacteria and protists according to University of

Cambridge's Alastair Crisp. Crisp's study added to the 17 already-reported genes integrated to the human genome by way of HGT. Aside from genes involved in metabolism, Crisp also found genes playing roles in immune responses, antimicrobial defense, antioxidant functions and protein modification. In addition to bacteria and protists, other groups like viruses and fungi have also been reported to contribute to the human genome through HGT that occurred along a time-span starting from the appearance of the common ancestor of Chordata up to the emergence of the common ancestor of primates.[68]

Humans have not been spared from HGT. For instance, the human immune system's origin has been associated with a horizontal gene transfer insertion event around 400 mya. However, HGT as a hallmark of the human genome has been refuted by some scientists. Steven Salzberg of Johns Hopkins University has claimed that the extraordinary conclusions made by Crisp and colleagues provided no extraordinary supporting evidence. Salzberg re-analyzed the various genes implicated in HGT into the human genome and arrived at a conclusion that not a single gene has been transferred in this manner. According to Salzberg, more mundane mechanisms are more likely to be behind the phenomenon observed by Crisp, noting that inheritance of genes by vertical descent has been established as the explanation for the presence of the vast majority of genes in the human genome.[69] It remains to be seen whether Crisp's assertions can be confirmed by more recent research and with the availability of newer tools and additional DNA sequence information. Nonetheless, HGT in metazoans is a probability given the prevalence of this process among viruses and bacteria. Are metazoans a special group of organisms such that they are exempted from HGT? It is quite difficult to think of any reason to support this.

We already know that virus genes are present in the human genome. As mentioned in the previous chapters, viral DNA makes up 8% of our genome. Ancient viruses were able to introduce their DNA into the genome of our ancestors around 100 mya and one of the evidences of these events is the presence of a protein called HEMO in the human fetus and placenta. A research group from Gustave Roussy Institute in France led by Thierry Heidmann has reported that the mysterious protein is expressed in pregnant women for no apparent reason.[70] What makes HEMO unusual is that it is produced by the fetus and in the placenta, not by the mother. Even more unusual is the source of the gene that encodes for this protein – a virus that infected the mammalian ancestors of humans. Heidmann suggested that HEMO proteins are some kind of signal for the mother from the fetus which tells the mother's immune system not to attack the fetus.

Other scientists offer more disturbing explanations. One such claim is that viral proteins are expressed in fetuses to help keep stem cells from losing their pluripotency or the cells' ability to turn into any tissue in the human body. As stem cells divide, they lose their pluripotency since they have to commit to becoming one of the potential cell types. Once they have managed to become a particular cell type, these cells usually turn off any viral genes they carry. By preventing stem cells from losing their pluripotency, viruses were able to coax the embryos to make more copies of the viruses and distribute these foreign invaders into the various parts of the human body. Viruses, however, could target eggs and sperm cells in particular because doing so increases the chances that the viruses are passed down as well to the next generation.

One of the viruses that have been regularly inherited in a small percentage of the human population is the human herpesvirus 6 or HHV-6. HHV-6 is the only human DNA herpesvirus that is capable of inserting its own genes into the human genome, allowing the virus to be routinely inherited by its host. According to a study by Alex Greninger and Louis Flamand from the Laval University in Quebec, the integration of the HHV-6's DNA into the human genome could have been accidental given the presence of repetitive DNA sequences in the virus genome just like what is observed in human chromosomes.[71] The research group has found that several of the HHV-6 genes were being actively expressed in a few human individuals, albeit with a low and sporadic expression. Nonetheless, the expression of selected genes was identified to be occurring in the brain, esophagus, adrenal gland and testes of some human hosts. The researchers also found that the HHV-6 genes inherited in the human genome help the body mount a stronger immune response against new invading viral proteins.

Another clue supporting the observation that the human genome can be used as a reference for genomes in other organisms is the existence of genes in some species of fish with human counterparts. This is only natural because humans and fish share a common ancestor that lived approximately 450 mya. The fact that humans and fish possess many similar genes is really not surprising since this is exactly how evolution works. However, what is remarkable is that most of these genes did not change dramatically since and are still performing the same roles in vertebrate anatomy and physiological development. Equally interesting is that there are still new genes in the human genomes being discovered through the analysis of fish genomes.

Take for example the case of pufferfish *Fugu rubripes*. An international team of researchers have reported the existence of 961 novel human genes matching those in the *Fugu* genome.[72] The human genome is approximately eight times larger than the *Fugu* genome although this discrepancy is attributed to the abundance of repetitive DNA in the former. However, 75% of the 31,000 genes in *Fugu* have been found to have equivalents in humans. It appears that the pufferfish has contributed a lot of its genome to ours! This also shows how much anatomical and physiological characteristics we humans share with other vertebrates. So, Neil Shubin was indeed spot-on when he said that we have an inner fish in our body. Shubin has shown in his book *Your Inner Fish* that our hands and heads somehow take after the fins and heads of ancient fish. It is really hard to deny the evidences when it comes to our evolutionary origin.

If our hands and head organization came from fish, then the chamber of our hearts can be traced back to turtles, lizards and other reptiles. That is according to researchers from the Gladstone Institute of Cardiovascular Disease in San Francisco, California. Benoit Bruneau's team has revealed the common genetic factor involved in the development of reptilian hearts. Their study has shown that the gene *Tbx5* holds the key to how evolution led to four-chambered hearts and warm-bloodedness. Bruneau has claimed that reptiles served as the link between the three-chambered hearts of amphibians and the four-chambered hearts of mammals, including ours. By studying mutations in the *Tbx5* gene, his team has found the protein it codes for is a transcription factor that, when disrupted, causes defects in the development of the ventricular septum or the wall of muscle between sections of a ventricle.[73] Knowing this, we can therefore say that we are partly reptilian deep within our hearts.

These are just some of the examples how humans share genes and morphological characteristics with other organisms on Earth, implying the common environments that all earthlings experienced during the billions of years of evolution of life. We can find more of such genes, organelles and organs that are present in humans but were actually from ancestral life forms, often with different original functions.

Another example are our cells' mitochondria which have been hypothesized to have descended from non-sulfur bacteria, incorporated into the prokaryotic or eukaryotic cytoplasm after surviving endocytosis. However, recent genomic studies are challenging these previously widely-acknowledged assumptions.[74] Nevertheless, no matter how tortuous the path that mitochondria took to reach us, the fact remains that this organelle is nothing but an important addition to human cells. We humans also have with us remnants of reptilian brain in the form of the basal ganglia. This primal brain has been known to control the self-preservation behavior of humans such as those related to eating, fighting, fleeing and reproducing.[75] In addition to the reptilian brain, we also have the paleomammalian (limbic system) brain and the neomammalian brain to complete the Triune Brain model developed by Paul D. Maclean.[76] For those not familiar with this information might find it highly interesting that our brain has a very intriguing history, too. The human lungs can likewise be traced back to our primitive ancestors. It has been reported that lungs did not really evolve from fish swim bladders, as Charles Darwin had suggested. Darwin recognized the homology between the two organs and suggested in his On the *Origin of Species* that the lungs of respiratory vertebrates could have been derived from a more primitive bladder. However, the lungs of humans and other land vertebrates are now being considered to have evolved from the primitive lung of a common ancestor – prehistoric *Polypterus* – we share with swim bladder-bearing fish.[77]

If we continue with the enumeration of the genes, organelles and organs that we share with or co-opted from our ancestors, we would be confronted with the undeniable fact that humans have virtually no original body part to speak of. This is to be expected because this is exactly how evolution works. It would not be surprising to find almost all of the genes in our genome to have duplicates in several other species. We can make new ones in the form of *de novo* genes and other mechanisms but the fate of these genes is a bit uncertain and their number is likely insignificant.

In other words, we humans are nothing but a living record of duplicated – albeit improved – genes of our ancestors. This is good news for the alien scientists because if they are able to obtain a single human specimen, they would be able to reverse engineer our body, study its modular parts, tease out all the relevant genes and generate a picture about the evolution of life and extrapolate this with Earth's geologic history.

Even if the human space explorers decide not to divulge the location of our home planet upon realizing that the alien civilization is not a friendly one (or even if the aliens found the whole human exploration crew deceased), these extraterrestrial sleuths would still be able to find Earth based on the evidences the human body would offer.

The first clue that could point to where the human explorers came from would be through the analysis of our circadian rhythm genes such as *BMAL1, NR1D1, BHLHE40, et cetera*. These are just a few of the major canonical clock genes that have been found to have circadian rhythmicity in their expression in peripheral human tissues and brains.[78] These can hypothetically lead alien scientists to the occurrence of day and night in the home planet of the human explorers. More importantly, they would likely find out about the existence of the 24-hour cycle of the human circadian rhythm.

They now have information about the amount of time that Earth takes to rotate with respect to the Sun. Next the alien scientists would probably have an idea about seasons and that these correspond to a planet's axial tilt or - possibly - revolution around its star (for planets with pronounced elliptical orbits). If they look close enough, then they would find seasonal genes in the human genome. Researchers from the Queensland Brain Institute have reported that expression of some genes involved in immune function have a 12-month seasonal cycle. The genes the research team found to have seasonality are those regulating blood cell count levels that involve red blood cells, neutrophils, monocytes and platelets.[79] This information could potentially provide alien scientists a rough estimate of the tilt of Earth's axis with respect to the Sun. In other planets with extreme orbits, seasonal genes could help aliens estimate the duration of a planet's orbit around its star.

So, they now have two important clues about Earth: that it rotates on its own axis and orbits a star (as expected) and that these take 24 hours and 12 months to complete, respectively. The alien scientists then would only need to search their databases for planets that rotate for 24 hours on their own axis and revolve around their stars for 12 months.

If the above information are not enough, they could narrow down their search further based on the probable size and mass of Earth. And how do they determine this? By looking at the density of bones of the human skeleton, researchers have found that we can stand gravity that is three to four times that of Earth.[50] This means that a bigger and denser planet could spell trouble for our bones, although it is also difficult to estimate the density of planets from afar. Nevertheless, alien scientists would probably be able to do the calculations and come up with a narrow range that covers Earth's radius which is 3,963 miles. Using these three educated guesses, the alien scientists would be able to obtain a few candidate planets for further cross-referencing or validation.

Having seeded Earth with the starting materials of life and/or accelerants of evolution, the alien scientists would likely have basic data about our planet and also know where to find us. If an entirely different advanced civilization happened to make contact with the human explorers or came upon their bodies, then their search for Earth would likely be less straightforward.

PART THREE:
Potential Co-conspirators

CHAPTER 1
A Place for Viruses

> *"We think of something that has four legs and wags its tail as being alive. We look at a rock and say it's not living. Yet when we get down to the no man's land of virus particles and replicating molecules, we are hard put to define what is living and what is non-living."*
>
> - Cyril Ponnamperuma
> *The Omni Interviews*, 1984

Viruses are a special group of 'living' entities on Earth. They are very familiar to humans due to the number of diseases they cause. But the more we look at their nature and origin, the more viruses appear to look strange and out of place in the grand scheme that is the evolution of life. Viruses are numerous and ubiquitous and, yet, they appear to lie outside the realm of where true living organisms exist. Perhaps viruses are more than just disease agents?

Viruses are often viewed as parasites because of their inability to reproduce on their own. In order to reproduce, viruses need to infect hosts and then manipulate the host's cellular machinery to produce more copies of the invader. Viruses carry genetic material in the form of RNA or DNA which can be single-stranded or double-stranded but are always enveloped by a capsid or protein coat. The genetic material and the protein coat make up the nucleocapsid but some viruses also have an additional lipid envelope. An entire virus particle is called a virion.[80]

The simplicity of viruses and apparent bareness of their structure have led many to view them as non-living entities. The most concrete evidence of this classification of viruses as non-living forms is their exclusion from the tree of life.[81] Viruses do not share any characteristic with cells. In other words, there is not a single gene shared between viruses and cells. Therefore, it is only logical that viruses should sit outside the tree of life.

Using genome analysis, evolutionary biologists have arranged all organisms on Earth to fit into a tree of life, directly representing the evolutionary relationships among earthlings. Every earthling can be placed under any of the three primary domains that include Archaea, Bacteria and Eukaryotes.[82] Prokaryotic cells or those lacking membrane-bound nuclei comprise the Bacteria domain and include Gram-positive and Gram-negative bacteria, cyanobacteria and mycoplasmas. Many Archaeans are extremophiles – such as methanogens, hyperthermophiles and extreme halophiles – although there are others that are not. Under Eukarya are protists, fungi, plants and animals. Viruses do not fit under any of these three domains although they all use nucleic acids as their genetic material. So what is the deal with viruses?

Thanks to Andre Lwoff, viruses have been considered nothing more than subcellular genetic parasites since 1957.[83] This view has been supported by the fact that viruses are unable to autonomously replicate and therefore need host cells' metabolic machinery to achieve this goal. Since viruses possess only the bare essentials – the nucleocapsid – to 'survive,' biologists have often considered viruses as mere parasitic genetic elements or "replicators."

As previously mentioned, all life forms undergo growth and morphogenesis as dictated by a blueprint that is their genetic material. Even microscopic bacteria are capable of morphogenesis as demonstrated by the development of swarmer cells into stalked cells in *Caulobacter crescentus* and flagellum biogenesis in *Helicobacter pylori*.[84] Morphogenesis in viruses is very basic which is equivalent to the formation of virion structures. The issue, however, is that viruses need the host cell's resources in order to 'grow' virions so this cannot be considered as autonomous growth.

Another characteristic of living organisms is reproduction or replication. The simplistic division of budding yeast *Saccharomyces cerevisiae* is a type of reproduction – asexual reproduction.[85] Viruses, on the other hand, go about reproduction by first attaching onto the cell surface of the host which engulfs the virus particle through a process called endocytosis. The virus then uncoats, releasing the genetic contents of the capsid. DNA or RNA strands then get inside the host's nucleus where these foreign genetic materials are replicated. In the host's cytoplasm, new viral proteins are synthesized to produce capsids that are needed for the assembly of new virions.[80]

All living organisms are also capable of auto-repair. We often see our skin and tissues gradually heal from wounds and injuries just like other animals. Plants also display this power to repair different parts, from roots to leaves. Even microscopic organisms are able to do so, as observed in the green alga *Chlamydomonas* through its regeneration of damaged flagella.[86] However, viruses do not have this mechanism, so this is a point against the argument for viruses as living organisms.

Viruses are also incapable of harnessing energy. Whereas plants and algae can harvest light energy to perform photosynthesis, viruses are fully dependent on other organisms for their energy sources. Viruses do not have enzymes for processing substrates in order to extract and consume their nutrients. Needing only to replicate and assemble virions, viruses can ensure their 'survival' by hijacking the metabolic machinery of their hosts.

Based on the abovementioned criteria, viruses cannot be considered as living organisms. Viruses can only satisfy one out of the four major characteristics of life. Thus, virus inclusion in the tree of life is a challenging proposition, especially when considering the following points: In addition to not having energy and carbon metabolism pathways, viruses also lack ancestral lineages. In fact, all viruses do not share even a single gene.[87] Organisms in the tree of life have lineages that point to their possible origin and evolution, typically supported by a number of genes that they share.

Viruses also lack structural continuity. One robust evidence indicating that all modern cells evolved from a common ancestor is the existence of a genetic membrane which viruses do not possess.

Some might argue that a few viruses display metabolic and translation genes, but these are mainly due to horizontal gene transfer from cells. This makes viruses appear as plain old gene robbers and nothing more.[88]

It is certainly intriguing to postulate, given the simplicity of viruses which triggers assumptions that they are ancient organisms, that they served as the missing link between the living and the non-living entities on Earth. Many maintain the belief that life on Earth evolved from simple to highly complex forms. However, evidences point to the fact that viruses could have followed the opposite path: regressive evolution. Instead of increasing the size and complexity of their genome, viruses tend to maintain a small genome to increase their replication rates. Therefore, viruses are not antique or primitive as their simplicity suggests. Their structure could just be a product of the evolutionary effects of parasitism.[89]

With all the revelations about viruses just mentioned, the question about their origin remains unsolved. Nevertheless, there are a few hypotheses that try to answer from where and how viruses evolved. Based on molecular studies, viruses are suggested to be ancient entities – so ancient that they could have existed before life diverged into the three domains of life we now have. Some paleovirologists even suggest that viruses could have played a role in the said divergence of life.[90]

It gets more complex when we consider the five hypotheses that have been forwarded to explain the place and role of viruses on Earth. According to the degeneracy hypothesis, viruses descended from small parasitic cells that targeted larger cells. Giant viruses such as the Mimivirus are structurally similar to that of a parasitic bacterium. Degeneracy hypothesis suggests that these small parasitic cells gradually lost many of their metabolic functions including their own replication machinery. The virus-first hypothesis suggests that viruses appeared on Earth before cells did and that viruses came from complex protein and nucleic acid molecules, suggesting that viruses had a role in the emergence of cellular life on Earth. The vagrancy hypothesis, meanwhile, contends that DNA or RNA from larger organisms served as the ancestors of viruses. These rogue genetic materials escaped from larger organisms and developed a mechanism for replication with the help of host cells.[89,90,91,92]

The abovementioned three hypotheses have their own limitations, however. For instance, the degeneracy hypothesis cannot account for the large discrepancy in structure and function between the hypothesized cellular parasites and modern viruses. The virus-first hypothesis has been criticized because it describes a virus that does not need a host cell for replication. Vagrancy hypothesis does not provide an explanation regarding the origin of viral capsids and other unique characteristics that are absent in cells. To address these challenges, newer hypotheses have been developed.

More recent hypotheses include the coevolution hypothesis and chimeric-origins hypothesis. According to the coevolution hypothesis, replicons thrived in hydrothermal vents or hot springs where food sources abound and where lipid-like molecules could have come from and served as the starting materials for the primitive form of capsids.[92] Chimeric-origins hypothesis, on the other hand, combines the virus-first and vagrancy hypotheses, proposing that viruses got their replication machinery from the primordial genetic pool while their structural proteins came from different host proteins as viruses evolved through time.[93]

These hypotheses highlight the fascinating status of viruses relative to the story of the evolution of life on Earth. Viruses are like an outcast, without origin and without relatives, at least for now. Therefore, it could be more convenient to just propose that their existence is a result of a direct introduction from outside the planet. For those espousing panspermia, viruses represent a suitable candidate that alien civilizations could have sent onto Earth to accelerate the evolutionary processes that have been started by the same sentient entities.

Again, we are assuming that our planet was an experiment that required occasional tweaking and monitoring by the extraterrestrial experimenters. If the alien visitors wanted to expedite their experiments, they could have resorted to applying some kind of catalyst to the evolutionary processes on Earth. Viruses appear like the appropriate tool for this purpose.

Through horizontal gene transfer (HGT), viruses could have speeded up the evolution of different species on this planet. We often wondered why it took unicellular organisms 3.5 billion years and another 2.5 billion years for multicellular life to emerge but it only required a little more than one billion years for all the biodiversity we see today to evolve? Could this discrepancy be due to the work of viruses?

A paper published by Aaron Vogan and Paul Higgs from the Origins Institute at McMaster University in Canada has reported the effect of HGT on genome evolution using a model. They found that a high rate of HGT is favorable when there is an increased rate of gene loss during genome replication. The authors add that HGT enables the rapid spread of new genes that result in larger and fitter genomes. However, the beneficial effects of HGT is only applicable during the early stages of evolution and not on modern prokaryotes, according to their model.[94]

It would be foolish to read too much from the results of the said research but knowing that there is a possibility that HGT can speed up the spread of genes in the early stages of prokaryotic evolution is an intriguing idea especially if we are looking even for a small chance that viruses could have been a catalyzing agent in the emergence of life on Earth. The odds may be low but the possibility is still there.

We can assume that any alien civilization that experimented with life on Earth would be aware that evolution is a predictable process and that organisms on Earth have modular storage and application of genetic information (both discussed in the following chapters), and could thus be expecting to observe the proliferation of a biodiverse life on the planet. To accelerate evolution, they could have introduced viruses by way of an asteroid, a comet or a spacecraft.

The technology to make all of these possible should not be that challenging for them. All they have to do is make sure that viruses survive in a controlled environment during the light years of space travel and that the virus load does not get incinerated during the vehicle's entry into the

Earth's atmosphere. The vehicle could have landed on any part of the vast ocean surface of our planet which should have softened the impact and provided a better environment for the release of the viral particles.

Why the ocean? Why not? The ocean is chockful of marine viruses. One liter of seawater is said to contain 100 billion viruses, most of them still unidentified. According to Curtis Suttle of the University of British Columbia, viruses play a huge role in carbon recycling in the world's oceans, turning over 150 gigatons of carbon annually, thus impacting the dynamics of Earth's systems. Viruses do this by infecting various microorganisms such as archaea, bacteria and microeukaryotes.[95]

Contrary to common belief, viruses are not really that very host specific. Recent studies suggest that viruses can infect a wide array of microbial species as well as other organisms that are distantly related. Viruses have been thus likened to zooplanktons and protists in their efficiency in killing microorganisms such as archaea, microeukaryotes and bacteria.[96]

Therefore, crash-landing onto the ocean is not only advantageous for viruses because of the reasons just mentioned. It is also favorable for them because water is also where life on Earth supposedly emerged from. Once delivered by the alien experimenters, viruses can work immediately with all the organisms found in the ancient lakes and oceans.

Animals and plants ventured into land around 470 mya.[97] Therefore, introduction of viruses into the world's oceans – for the purposes of accelerating evolution – could have been done before the Cambrian Period although the explosion in biodiversity on Earth occurred around 1 bya, so it would make more sense if the alien civilization sent their virus payload between 2.5 bya and 1 bya.

It would have been difficult for viruses to infect as many bacteria, algae and animals on land since air is not an appropriate medium for viruses unlike water. Eventually, viruses would infect terrestrial animals but they could have needed vectors such as insects or they would have to be highly infectious to do so.

Junk DNA

> *"There were long stretches of DNA in between genes that didn't seem to be doing very much; some even referred to these as "junk DNA," though a certain amount of hubris was required for anyone to call any part of the genome "junk," given our level of ignorance."*
>
> - Francis S. Collins
> *The Language of God: A Scientist Presents Evidence for Belief*, 2006

Is junk DNA really junk? This is probably one of the most important questions that have boggled the minds of geneticists and evolutionary biologists around the world. These scientists (used to) use the term 'junk DNA' to refer to a region of an organism's genome that is noncoding, as opposed to coding regions. In other words, junk DNA is DNA lacking instructions for the synthesis of any protein that a cell needs. The current understanding of this type of DNA has been rapidly changing in recent years but the name has unfortunately stayed. It was Susumu Ohno who coined the term *junk* DNA in 1972.[98] Back then, the name proved quite catchy but it ended up discouraging many researchers from devoting time and resources to reveal the mystery surrounding this noncoding genetic material for almost two decades, according to Wojciech Makalowski of Pennsylvania State University. He adds that it was only in the early 1990s that scientists began giving more attention to noncoding DNA. Now, noncoding DNA are no longer considered entirely junk DNA. In fact, researchers have identified several roles that noncoding regions play.

Recent studies have shown that noncoding regions have essential functions, too, such as the transcription of RNA molecules that regulate gene expression by turning genes on or off. The Encyclopedia of DNA elements (ENCODE) project has revealed that a large proportion of noncoding DNA has actual roles in gene regulation. The project's findings indicate that modifications in gene activity regulation that is made possible by elements in the noncoding sequences could lead to protein synthesis disruption as well as other cellular processes thus leading to disorders. In other words, even if these regions do not code for proteins, they are not completely useless because they perform other functions and are therefore not junk.[99] However, there have been plenty of criticisms [100,101] when the researchers from the ENCODE project declared that there is "no more junk DNA."

One specific role is holding the genome together, as Yukiko Yamashita and colleagues at the University of Michigan Life Sciences have reported.[102] They studied pericentromeric satellite DNA regions that come as long, repetitive sequences and a protein that bind to satellite DNA called D1. The research team found that D1 and satellite DNA are important in packaging chromosomes together in one place thus preventing them from escaping the confines of the nucleus. Therefore, the researchers noted, noncoding DNA regions are important for cellular survival in all species that stash their DNA into the nucleus – that include us humans.

The most interesting of these recently discovered functions of the noncoding DNA, however, is interacting with its genomic environment and enhancing the potential of organisms to evolve. As discussed earlier, not all genes we find in organisms today evolved from ancient or older genes. Orphan or *de novo* genes are examples of genes that organisms build 'on the fly.' A gene that appeared to have been built from scratch is one that codes for an antifreeze protein in fish. Helle Tessand Baalsrud from the University of Oslo discovered this gene in the Atlantic cod *Gadus morhua*.[103] The genes *afgps*, that code for the antifreeze protein in cods, are not homologous with any other gene and are thus considered orphan genes. Baalsrud found that *afgps* came from completely non-coding DNA, giving fish antifreeze function as a probable mechanism for freeze avoidance in codfishes and notothenioids during the time when Earth went through periods of freezing cycles, approximately 800 mya.

These surprising findings about junk DNA demonstrate the hidden power of seemingly useless genes in 'recording' events on Earth – if we go by the logic that genes can be associated with a specific environmental cue or condition, as has been discussed previously. The evolution of the antifreeze gene is a fine example of how particular events on Earth can leave a mark on an organism's genome. What triggered the birth of this de novo gene is the series of freezing cycles. Apparently, without such extreme conditions, the gene would not have been created out of the noncoding DNA.

The noncoding region of an organism, in this regard, works like a library of templates for genes. We might as well look at the noncoding part of the genome as an organism's DIY kit "under its sleeve." Considering that one out of every ten genes in any genome is a *de novo* gene, it means that this is one useful kind of kit for those species looking to evolve as the planet changes – which is basically true for all of us earthlings. So, it seems that life forms wanted to come prepared for any possibility and it is therefore not surprising that some species brought enormous sizes of noncoding DNA with them. If organisms are in it for the long run in terms of evolution, then they better do so, don't they?

Now, the important question is, what does *de novo* gene-birth-through-noncoding-DNA implies? For evolution in the big picture, it would probably be not too big of a deal. Hey, evolution is about constant change after all! However, if we look at evolution as a measure of what can be discoverable about Earth, then the genetic materials of organisms, including noncoding regions, can serve as a way to indirectly estimate all the possible genetic functions that can be created. In other words, the genomes of organisms are like computer memories with petabytes of storage capacity. Therefore, alien scientists could have known and could have placed a lot of importance on this characteristic of genomes. In this regard, junk DNA can no longer be seen as accidental components of genomes. Instead, it would appear that junk DNA have been integrated into primitive genomes or allowed to propagate and be inherited to act as some kind of "extra memory space" in addition to the other regulatory and structural functions that have been elucidated by recent research on noncoding DNA. The idea is that the designer of primitive genomes could have added junk DNA with an intention to harvest fully expended genomes far into the future. Isn't that what we would have done if we were alien scientists and we consider genomes as organic data recorders?

CHAPTER 3

Modularity of Biological Systems

> *"As the human fetus develops, its changing form seems to retrace the whole of human evolution from the time we were cosmic dust to the time we were single-celled organisms in the primordial sea to the time we were four-legged, land-dwelling reptiles and beyond, to our current status as largebrained, bipedal mammals. Thus, humans seem to be the sum total of experience since the beginning of the cosmos."*
>
> - Jonas Salk
> *Omni*, 1982

Modularity of biological systems is an important phenomenon in biology. This stems from the fact that modularity is ubiquitous in biological systems. Nature is full of examples of independent components that can stand alone in terms of their morphology or function – referred to as biological modules – enabling organismal growth, development and evolution. An oft-used story for demonstrating modularity is the classic parable of

Hora and Tempus by H.A. Simon. It has been quoted as an introduction in an article discussing the emergence of modularity in biological systems before[104] but it is worth repeating the story here because it helps illustrate the concept clearly:

> "There once were two watchmakers, named Hora and Tempus, who manufactured very fine watches. Both of them were highly regarded, and the phones in their workshops rang frequently — new customers were constantly calling them. However, Hora prospered, while Tempus became poorer and poorer and finally lost his shop. What was the reason?
>
> The watches the men made consisted of about 1,000 parts each. Tempus had so constructed his that if he had one partly assembled and had to put it down — to answer the phone say— it immediately fell to pieces and had to be reassembled from the elements. The better the customers liked his watches, the more they phoned him, the more difficult it became for him to find enough uninterrupted time to finish a watch.
>
> The watches that Hora made were no less complex than those of Tempus. But he had designed them so that he could put together subassemblies of about ten elements each. Ten of these subassemblies, again, could be put together into a larger subassembly; and a system of ten of the latter sub-assemblies constituted the whole watch. Hence, when Hora had to put down a partly assembled watch in order to

answer the phone, he lost only a small part of his work, and he assembled his watches in only a fraction of the man-hours it took Tempus."

Simply put, modularity implies that across various scales, biological systems can be broken down into smaller components or systems, with minimal interruptions between them. In other words, these biological systems have intensely interacting parts but are autonomous with respect to other systems. A biological module, therefore, is a component of a cell, tissue, system or organism that is embedded in a particular process but is relatively autonomous from other components of a cell, tissue, system or organism.[105] Fine examples of modules include bacterial plasmids and endosymbionts. Plasmids, which are extra-chromosomal mobile elements, often mediate horizontal gene transfer. Naturally, plasmids confer bacteria such as *Bacillus thuringiensis*, *Bacillus cereus* and *Bacillus anthracis* with virulence factors, pathogenicity and genes involved in catabolism, for instance.[106] Endosymbionts, on the other hand, are organisms that thrive inside other organisms. Many endosymbionts cannot survive outside their hosts. Examples of endosymbionts include Wolbachia, Wigglesworthia and Buchnera; these three organisms are proteobacterial endosymbionts of insects.

The advantage of modularity in biological systems is that it enables changes in one module without impacting other modules or the rest of the system. This makes modularity of developmental processes a significant factor in promoting evolution.

Evolvability is the potential of a system to generate beneficial variation through the help of random genetic modification.[107] This property of living things should make modularity a significant part of discussion when it comes to the concept of directed panspermia.

Genetic or metabolic modules could easily be propagated on Earth. These modules can easily evolve given enough time and the right conditions. This would be a logical strategy if an alien civilization knows about this phenomenon about life forms. In fact, they would be able to predict how and when life would evolve on Earth if they understand how modular biological systems work. Therefore, modularity could be a basis – if not *the* rationale – of an interstellar evolution experiment.

Lucio Vinicius offers an informative look into the importance of modularity in the context of evolution. His book, *Modular Evolution: How Natural Selection Produces Biological Complexity*, provides ample evidence to the author's claim that 'modularity transfer' is behind the complexity of life on Earth.[108] According to Vinicius, modular phenotypes such as proteins, cells and behaviors underwent evolution to become novel modular information carriers in the form of regulatory proteins, neural cells and words to give rise to highly complex biological systems we now find in nature. He also considers humans as the pinnacle of biological complexity, thanks to modular evolution.

It is hard to argue against Vinicius's assertions that humans are the prime achievement of modularity of biological systems and evolution. Indeed, life – from the smallest microbes to the largest mammals – appears to be composed of biochemical Lego blocks that represent biological modules. Many scientists consider humans as the finest iteration of evolutionary sophistication. However, agreeing with this claim would mean that the story ends with the evolution of humans, who have also invented non-biological evolution as Vinicius has noted. So, is this it? After becoming the centerpiece of this planet's living world, are humans supposed to just maintain this position? Or, are humans going to wait or enable the rise of other animals to reach their own potential?

For those on Earth, the abovementioned supposition should be enough to explain the existence of things as they are. But given the vastness of the universe and the possibility that a number of advanced civilizations could be thriving on other planets, it would be a kind of shortsightedness for anyone to think that evolution is limited to this planet. Modularity of biological entities should be found in other planets as well. It is even quite sensible to say that, along with possessing genetic materials such as DNA and RNA, modularity could be considered as hallmarks of living things in the whole universe.

Modular units should be ubiquitous in the universe; it is the way that many things are efficiently organized, albeit temporarily: atoms, molecules, cells, tissues, organs, bodies, planets, star systems and galaxies. If we follow the pattern, the universe could also turn out to be a modular part of a larger system. This is why it would be quite easy to comprehend if astrophysicists should choose to assume that our universe is a part of an unimaginably large super-universal system. But this is beyond the scope of this book and I have no business speculating on the topic. What we are only demonstrating here is that modular units increase in size and complexity as we go along or while zooming out of the picture. In the same way, a human looks like a modular system composed of smaller and smaller modular systems and could also be a part of a larger, unseen modular system we are yet to discover.

If an advanced civilization is aware of the ubiquity of life, then it would be easier for us to understand why aliens seeding life and experimenting with modular biological units on other planets is a logical or even an attractive plan. For them, life could be just like a recombinant DNA experiment – a virtual molecular Lego exercise!

CHAPTER 4

Predictability of Evolution

"Replay the tape a million times... and I doubt that anything like Homo sapiens would ever evolve again,"

- Stephen J. Gould
Wonderful Life: The Burgess Shale and the Nature of History, 1989

What would an ancient civilization be asking if they are going to experiment with evolution by seeding a biologically-empty planet like Earth? Why do it in the first place? This is like asking one of the most important questions there is in the field of biology. For many, that question could be, "what would we get if we evolve life from the very start?" Would we see the same level of biodiversity? Would there be highly intelligent organisms among the evolved life forms?

Stephen J. Gould once said that if we rewind time and let evolution on Earth happen again, there might not be a 'we' the next time around.[109] Gould claimed that *Homo sapiens* might not re-evolve because life would take an entirely different path from the one that led to what we have now. He believed that human evolution was a one-off event that will not happen again even if the tape of life was replayed a million times. Gould attributed this rarity to chance events having an enormous role in evolution such as mass extinction events, as discussed previously. These include bolide impacts and supervolcano eruptions.

In addition to these environmental calamities are molecular chance events in the form of mutations in the genomes of organisms. One gene mutation can theoretically alter the survival capabilities of one population. A set of mutations can therefore cause the evolutionary tree of life to drastically change especially if the affected genes happened to be the critical ones for the evolution of a group of species. For example, the set of genes that were instrumental in the formation of limb-like limbs in *tiktaalik*, can be considered crucial for *Tiktaalik roseae*'s ability to propel itself on land and raise its head to look for food in a terrestrial environment. If one of the limb genes suffered a mutation, it might be safe to say that *tiktaalik* could have stayed in the water and prevented its predecessors from conquering land. In that case, we can say that Gould was probably correct.

However, removing *tiktaalik* from the picture does not mean that animals will not be able to eventually occupy land. There could be other organisms that might have served as the missing link between water-dwelling and land-dwelling animals. The path might be different but the end-result should be the same. Given enough time, evolution of a particular mechanism – such as walking on land – can be expected to occur as various species search for ways to improve their chances of survival. Still, Gould was right, it seems!

There are indications, though, that evolution is predictable after all. By predictable we mean evolution producing characteristics in organisms that have been observed before. For example, researchers have demonstrated the evolution of the same traits after numerous generations in several populations of a bacterium. Richard Lenski of Department of Microbiology and Molecular Genetics at Michigan State University has been maintaining 12 populations of *Escherichia coli*, which were initially identical, since 1988. These populations of *E. coli* have been continuously evolving which means they have undergone more than 65,000 generations as of 2017.

To put that into context, there have only been 1,000 to 10,000 generations of *Homo sapiens*. That is how young the human population is compared to the *E. coli* Lenski and colleagues are studying. So, what have they found? The research team observed similar patterns of rapid enhancement in fitness in all the 12 populations of the bacteria. In addition, they also noted that all the *E. coli* populations are displaying larger cells and quicker growth rates.[110] If this finding is not indicative of the repeatability of evolution, I am not sure what is.

Another study demonstrating the repeatability of evolution involves insects. Anurag Agrawal and his team investigated 18 species of insects from four orders regarding their adaptation to cardenolides – potent toxins found in plants that the insects feed on. The 18 insect species included moths, butterflies, flies, beetles and true bugs that forage on cardenolide-containing plants such as foxglove and milkweed.

Agrawal's team zeroed in on the sodium pump (Na,K-ATPase) gene. A mutation, called N122H, in this gene confers resistance to insects against cardenolides. Interestingly, the research team found that the N122H mutation is present in all of the four insect orders studied.[111] This could not be pure coincidence alone. Indeed, different species of insects respond similarly to a common environmental cue signifying a version of convergent molecular evolution.

There are other studies pointing to the same conclusion but we do not need to discuss them here. However, do these two and other snippets of evolution prove that Gould was, in fact, wrong? It is difficult to say so but maybe now we can assume that if the tape of life is rerun we are probably going to see some familiar faces along the way. This is enough evidence for those looking for indications that evolution is predictable especially as a rationale for an alien civilization's experiment proposed here.

In other words, "will evolution eventually lead to a sentient species?" could be a credible enough question for any self-respecting alien scientist to pursue and this might be the reason that we are on this planet now.

PART FOUR:
A New Picture

A NEW PICTURE

CHAPTER 1

Synthesis

> *"A popular cliché in philosophy says that science is pure analysis or reductionism, like taking the rainbow to pieces; and art is pure synthesis, putting the rainbow together. This is not so. All imagination begins by analyzing nature."*
>
> - Jacob Bronowski
> *The Ascent of Man*, 1973

So, what have we learned thus far? In *Part One* we were re-introduced to the idea of directed panspermia and that evolution of life on Earth could have been ignited by an alien race looking to confirm a hypothesis on the origin and evolvability of intelligent species. This author does not claim that this is what happened but the following points can be considered as potential explanations:

Life on Earth could have been provided the precursor components or evolution might have been accelerated using viruses by an alien civilization. As discussed in *Part Two*, potential mechanisms that could have been exploited for this method are the evolution of intelligence and the evolution of curiosity in order to produce a candidate species - humans. Along the way, this candidate species has been designed to attain the smarts and wandering spirit while recording every detail that can be recorded about its home planet. Thus, the genome of *Homo sapiens* became a repository of a humongous amount of information not just about extant microbes and fauna but also about the past conditions on Earth. Humans, in short, are a walking compendium of the biochemical blueprints of their predecessors from the prehistoric cells to invertebrates, fishes, reptiles and mammals whose parts and behavior have been adopted and co-opted by the most evolved species thus far: our species.

If we take the point-of-view of the alien experimenters, the human species is the ideal medium for extracting and importing organic data from Earth towards the alien's home planet. Human explorers can be the vehicle, storage medium and data in one! All they have to do is to wait - if they have enough time - for humans to send signals about our existence and location or come to their planets inside spacecraft the humans themselves made. How convenient.

Then in *Part Three*, we are reminded of the phenomena related to evolution that could have helped the agenda of alien scientists in the aforementioned endeavor. We have already pointed out the potential of viruses to speed up evolution through HGT. Next, we learned that junk DNA may not be useless after all. In fact, noncoding DNA could serve as extra memory space for all the new information that humans can gather in the future via the generation of *de novo* genes. Noncoding DNA is found in all organisms which means that the potential for evolution can be sort of limitless in this sense. For the purpose of documenting the environment, the human genome can take in more information but could also be ripe enough now according to the alien scientist's liking.

We also reviewed the modularity of biological systems and found this to be an advantageous trait on the part of alien scientists planning to experiment on life on Earth. Modular units are ubiquitous in nature and knowing this can help them design how evolution should progress on this planet. Alternatively, modularity could have been an expected effect of seeding the precursors of life in the primordial soup. The alien scientists should be familiar with modularity since they, in all likelihood, could be a product of modularity themselves.

Finally, we recognized the predictability of evolution which could be both the question and rationale that the alien experimenters had in mind in simulating evolution on Earth. That organismal traits can be evolved repeatedly says a lot about the familiarity of the story of life on Earth and in other planets, if there turned out to be any. This realization should have served as enough impetus for anyone wanting to know how special (or not) their kind is to rewind and replay the tape of life.

In short, if an alien species would like to know more about its own origin, following the above strategy is a good way to do it. There could be other ways but it is highly likely that they would end up settling for this option. Why? Because what have been discussed here are the universal properties of life and evolution not just on Earth but maybe even on other planets, too.

Some scientists believe that the genetic material in alien life could differ remarkably from those on Earth. Instead of ACGT - for adenine (A), cytosine (C), guanine (G), or thymine (T) - other units could be used by alien as their 'DNA.' Instead of phosphorus, extraterrestrial life could exploit other elements abundant on their planet for their 'DNA' backbone. However, it is doubtful that the general structure fashioned after the double helix could be ignored by evolution in these distant worlds. The DNA helix is a highly efficient configuration that should be favored by innovating primitive life forms. The same thing can be said about the structure and replication mechanisms of chromosomes. It would be interesting to see what other storage techniques can be developed by alien life for their genetic information.

The existence of genetic change instigators in the form of viruses is likewise an obvious strategy, such that any alien scientist can easily come up with it. If one wants genetic material to mix and match among the experimental specimen, then small, efficient and infectious genetic vectors are needed.

Viruses are convenient tools for such a purpose and if there comes a time when the evolutionary position of viruses in the tree of life is finally settled, then the suggestion forwarded in this book regarding the possibility of viruses as nothing but introduced catalysts of genetic change should be summarily discarded and forgotten.

The same argument might be made regarding the suggestion that junk DNA could be extra memory storage that organisms can use for making new genes as they evolve and also serve as repository of environmental information. But, if a scientist wants to design genetic material that enables evolution while enabling the creation of new genes and documentation of life's history, then exploiting noncoding DNA is probably an ingenious way to do it. It just makes sense!

Similarly, it might be difficult to make evolution work without modularity. Modularity should be the defining feature of any organic life in order to achieve rapid evolution. As the parable of Horus and Tempus shows, following a non-modular strategy to make a clock can be painstakingly slow. Therefore, if one happens to land on an inhabited planet, aside from having genetic material, the observer will most likely find that life there are modular biological systems analogous to those on Earth. How can we say this? Because not following a modular strategy just sounds stupid and nature is not stupid but rather highly intelligent in its tendency to find the closest paths and connections - not to mention making elegant biological designs.

And the most important testable hypothesis that any alien experimenter on evolution should make the cornerstone of their research is predictability of the whole process. The goal is to be able to replicate the origin of life and emergence of sentient species. The alien scientist would likely not proceed with inoculating Earth with the starting materials of life without first suspecting that a biodiverse planet would result from that action. It should be an obvious question for any self-respecting biologist. For all we know, we might only exist as a result of a graduate student's research. So much for the vaunted uniqueness of our species!

A NEW PICTURE

CHAPTER 2

A Sublime Display of Altruism

> *"We are survival machines - robot vehicles blindly programmed to preserve the selfish molecules known as genes. This is a truth which still fills me with astonishment."*
> — Richard Dawkins
> *The Selfish Gene*, 1976

Richard Dawkins is known for coining the term 'selfish gene' which he explained in his book of the same title.[112] The gene-centered view of evolution promoted by Dawkins has been well-received by many evolutionary biologists and it has revolutionized biology. According to this concept, gene selfishness typically leads to selfishness in individual behavior. However, it has also been proven that animals sometimes demonstrate a limited form of altruism which can also help genes to achieve their selfish goals. In other words, organisms can behave altruistically thus negatively affecting their own interests, such as safety and reproductive ability, but still improve their chances of spreading their genes through the related organisms.

In light of the proposal mentioned here wherein humans serve as a reporter species and a representative to make contact with the alien experimenters on behalf of all the life forms on Earth, it can be said that the selfish gene theory could still holds true in such an event. Humans could act altruistically by getting in touch with the alien civilization, a decision which can severely affect the survival of humans on Earth. What might be considered as detrimental to the human race - such as obliteration of our civilization - could end up perpetuating the human genome eventually. Alien researchers could harvest our DNA for cloning later on or they could take founder populations (amicably, hopefully) from Earth for propagation or breeding on their planet. Either way, our genes could gain the potential of being spread on other planets as well.

What the above scenario indicates is that the decision of human explorers to search for extraterrestrials can be considered as a natural tendency and is therefore not entirely unexpected of our species. The fate of the human civilization, however, could depend on the motivation or ultimate plan of the alien civilization. If they decide to let humans exist on Earth, which is highly likely, then it means that there is really no cause for concern for us.

So, are we really headed down this road no matter what we do? Is it our fate to meet up with life forms from another galaxy? In a way, the answer could be yes. We have been looking at other planets to colonize since we found out that space travel is possible. We are aware that a pandemic, bolide impact, nuclear winter or climate change could render Earth inhabitable in an instant. Recently, NASA and other space agencies have been developing plans to make Mars suitable for human life. These are signs that the motivation is there and that human interplanetary migration happening is really just a matter of when, not why. Our genes are looking for ways to perpetuate themselves and it does not matter how this is achieved. After all, people are just vehicles of our own genome, according to Dawkins. Although the Earth-based existence of the human genes has been secured for now, the vulnerability of this planet is a clear indication that our genes are not completely safe from extinction in the future.

Now, should we aggressively look for alien life outside Earth? I think this is a given. It is in our genes to be curious and there is no need to suppress this instinct. Curiosity has benefited us in many ways and should continue to do so for our species. However, we might find it an advantage to rethink our strategy. We should be a bit cautious when we broadcast our location to other species. This might seem to echo what the renowned physicist Stephen Hawking had said, warning us of the danger of reaching out to aliens. Indeed, based from the ideas discussed in this book it seems that we should be prudent with space exploration especially for the purpose of contacting extraterrestrials. We should continue actively observing the universe for signs of life but let us make sure that we are going to discover them first. This way, we can be sure of their capabilities and motives before we 'speak' to them.

CHAPTER 3

Implications

> "'Why go to the stars? Because we are the descendants of those primates who chose to look over the next hill. Because we won't survive here indefinitely. Because the stars are there, beckoning with fresh horizons.' - James and Gregory Benford"
>
> - Michio Kaku
> *The Future of Humanity*, 2018

Knowing the possibility - no matter how remote - that we could be a product of an intergalactic experiment might trigger a feeling of uselessness in some people. On the contrary, this idea can be met with a degree of positivism: that the human evolutionary story is still unfinished. Achieving dominance over the other animals on Earth often gives us a sense of having accomplished our "evolutionary duty."

At the pinnacle of this planet's biodiversity, the vantage point offers us some level of security. Aside from occasional pandemics and natural calamities, it seems that we have mastered our fate as a species. We can transform our environment according to our needs and we can make vaccines to fight infectious diseases. We are aware of the threat of bolide impacts thus we can plan ahead of such events. At this stage, our fight is no longer competing for food and space against other animals. Now, our species is mainly concerned with maintaining our apex position. We only need to make sure our civilization is stable to ensure our survival. That's it.

But this is not the end. There are more in store for our species if we only chose to evolve and maximize our potential. The prospect of seeing what our species would look like after 50,000 or more generations or after colonizing other planets is surely interesting. What, we might ask, other species in the *Homo* genus could emerge in the future?

According to Peter Ward, the former project leader of the node of NASA Astrobiology Institute at the University of Washington, there are three routes that human evolution could take. The first route is called *stasis* wherein humans do not drastically change except for having more races. The second route is *speciation* which is basically the emergence of new human species on Earth or in another planet. The third and last route is *symbiosis with machines* or the merging of humans and machines, creating a collective intelligence that could be entirely different from the human form that we are familiar with.[113]

It would appear that the second route - speciation - is the most dramatic possible outcome for human evolution among the three options. The possibilities would be endless if humans decide to colonize other planets. On this path, the story of the human species - selfish gene, 'candidate' or 'reporter' genes, junk DNA, viruses, the modularity of biological systems and the predictability of evolution - continues.

On our part, we can take consolation from the fact that we are the caretakers or enablers of these biological phenomena that could potentially radiate from Earth. This could be our noble purpose as a species. We might not have the opportunity to become founder individuals on other planets but our supporting role should be just as important.

There are also lessons that can be gleaned from the concepts discussed here for application in our daily lives. One takeaway is that evolution is not a race and reaching the pinnacle does not give us humans the right to hold dominion over other species. We are the product of a collaborative effort involving billions of years of trial-and-error by countless species from unicellular forms to the vertebrates. This should remind us of the importance of small role-players in our work and in our personal lives. Our jobs and achievements should come easy if we take care of the bits and pieces, the cogs that make the machines work.

Being the most sophisticated species on this planet should also push us to do more to preserve the biodiversity all around us. We may not directly benefit from the existence of some exotic organisms but these could be important links in the evolution of some taxonomic groups. Awareness of this fact should intensify conservation campaigns around the world. More importantly, we should find a number of techniques to preserve the genetic information (spores, seeds, eggs, sperms, etc.) of various organisms that are on the brink of extinction.

The second lesson is the importance of modularity in our lives. Modular units are not only found in living things but could also arise in inanimate objects and in abstract ideas such as those that govern societies. We should construct our world and interactions following modular examples to increase efficiency and prevent any sudden collapse of systems. This should also enable the fluidity of necessary changes as well as rapid development in the workplace, in our homes and in society as a whole. We could experience significant improvement in our lives if we only make modular concepts a priority in designing our surroundings.

In summary, if we desire an efficient, dynamic and adaptive system, we should look at our own evolutionary origins and learn from the mechanisms that our ancestors have used to thrive on this planet for billions of years. The key to our success in life lies in our recognizing the numerous tools that are already built-in to our being. The human body can still be full of surprises and more biological and evolutionary insights from our species should become common knowledge in the coming years as we continue to research the secrets of the cell and our genome.

REFERENCES

1. Crick, F.H.C. and L.E. Orgel (1973). Directed panspermia. *Icarus* (19)3: 341-346

2. Bialy, S. and A. Loeb (2018). Could solar radiation pressure explain 'Oumuamua's peculiar acceleration?' *The Astrophysical Journal* 868(1). DOI=10.3847/2041-8213/aaeda8

3. Gilbert, E.A., Barclay, T., Schlieder, J.E., Quintana, E.V., Hord, B.J. et al. (2020). The first habitable zone earth-sized planet from TESS. I: Validation of the TOI-700 system. *arXiv*:2001.00952

4. Altwegg, K., Balsiger, H., Bar-Nun, A., Berthelier, J.-J., Bieler, A. et al. (2016). Prebiotic chemicals – amino acid and phosphorus – in the coma of comet 67P/Churyumov-Gerasimenko. *Science* (2)5: e1600285

5. Ball, J. (1973). The zoo hypothesis. *Icarus* 19(3): 347-349

6. Villamoare, B., Kimbel, W.H., Seyoum, C., Campisano, C.J., DiMaggio, E.N. et al. (2015). Early Homo at 2.8 Ma from Ledi-Geraru, Afar, Ethiopia. *Science* 347(6228): 1352-1355

7. Lillie, R.S. (1922). Growth in living and non-living systems. *The Scientific Monthly* 14: 113-130

8. LaBar, T., Adami, C. and A. Hintze (2015). Does self-replication imply evolvability? *Proceedings of the European Conference on Artificial Life* 595-602

9. Tang, S.K.Y. and W.F. Marshall (2017). Self-repairing cells. *Science* 356(6342): 1022-1025.

10. Moore, A. (2012). Life defined. *BioEssays* 34: 253-254

11. Lane, N., Allen, J.F. and W. Martin (2010). How did LUCA make a living? Chemiosmosis in the origin of life. *BioEssays* 32:271-80

12. Haldane, J.B.S. (1929). The origin of life. *The Rationalist Annual* 148: 3–10

13. Robertson, M.P. and G.F. Joyce (2012). The origins of the RNA world. Cold Spring Harbor Perspectives in Biology 4(5): a003608

14. Lazcano, A. (2016). Alexandr I. Oparin and the origin of life: A historical reassessment of the heterotrophic theory. *Journal of Molecular Evolution* 5-6: 214-222

15. Steigman, G. (2001). *Encyclopedia of astronomy and astrophysics.* Bristol, UK: Nature Publishing Group/Institute of Physics Publishing

16. Zeng, L., Jacobsen, S.B., Sasselov, D.D., Petaev, M.I., Vanderburg, A. et al. (2019). *PNAS* 116(20): 9723-9728

17. Shubin, N. (2013). *The universe within: The deep history of the human body.* New York, United States: Knopf Doubleday Publishing Group

18. Chaplin, M.F. (2001).Water: Its importance to life. *Biochemistry and Molecular Biology Education* 29: 54-59

19. Bellissent-Funel, M.C., Hassanali, A. Havenith, M., Henchman, R., Pohl, P. (2016). Water determines the structure and dynamics of proteins. *Chemical Reviews* 116(13): 7673-7697

20. Debès, C., Wang, M., Caetano-Anollés, G., and F. Gräter (2013). Evolutionary optimization of protein folding. *PLoS Computational Biology* 9(1), e1002861

21. Jordan, S.F., Rammu, H., Zheludev, I.N., Hartley, A.M., Marechal, A. and N. Lane (2019). Promotion of protocell self-assembly from mixed amphiphiles at the origin of life. *Nature Ecology & Evolution* 3: 1705-1714

22. Knauth, L.P. (1998). Salinity history of the Earth's early ocean. *Nature* 395: 554-555

23. Tosca, N.J., Knoll, A.H. and S.M. McLennan. Water activity and the challenge for life on early Mars. *Science* 320(5880): 1204

24. Landis, G.A. (2001). Martian water: Are there extant halobacteria on Mars? *Astrobiology* 1(2): 161-4

25. Stöffler, D., Horneck, G., Ott, S., Hornemann, U. and Cockell (2007). Experimental evidence for the potential impact ejection of viable microorganisms from Mars and Mars-like planets. *Icarus* 186(2): 585-588

26. Knoll, A.H. and M.A. Nowak (2017). The timetable of evolution. *Science Advances* 3(5): e1603076

27. Jablonski, D. (2004). Extinction: Past and present. *Nature* 427(589)

28. Pope, K.O., D'Hondt, S.L. and C.R. Marshall (1998). Meteorite impact and the mass extinction of species at the Cretaceous/Tertiary boundary. *PNAS* 95(19): 11028-11029

29. Alvarez, L.W., Alvarez, W., Asaro, F. and H.V. Michel (1980). Extraterrestrial cause for the Cretaceous-Tertiary Extinction. *Science* 208(4448): 1095-1108

30. Robertson, D.S., McKenna, M.C., Toon, O.B.., Hope, S., J.A. Lillegraven (2004). Survival in the first hours of the Cenozoic. *GSA Bulletin.* 116(5–6): 760–768

31. Pavé, A., Hervé, J.C. and C. Schmidt-Lainé (2002). Mass extinctions, biodiversity explosions and ecological niches. *Comptes Rendus Biologies* 325(7):755-765

32. Melián, C.J., Alonso, D., Allesina, S., Condit, R.S. and R.S. Etienne (2012). *PLoS Computational Biology* 8(3): e1002414

33. Pigot, A.L., Sheard, C., Miller, E.T., Bregman, T.P., Freeman, B.G. et al. (2020). Macroevolutionary convergence connects morphological form to ecological function in birds. *Nature Ecology & Evolution* 4: 230-239

34. Parker, E. T., Cleaves, J. H., Burton, A. S., Glavin, D. P., Dworkin, J. P., Zhou, M., Bada, J. L., and F.M. Fernández (2014). Conducting Miller-Urey experiments. *JoVE* 83: e51039

35. Lineweaver, C.H. (2009). Paleontological tests: Human-like intelligence is not a convergent feature of evolution. *arXiv* DOI:10.1007/978-1-4020-8837-7_17

36. Shubin, N. (2008). *Your inner fish: A journey into the 3.5-billion-year history of the human body.* New York, United States: Knopf Doubleday Publishing Group

37. Hesse, B. E. and B. Potter (2004). A behavioral look at the training of Alex: A review of Pepperberg's the Alex studies: Cognitive and communicative abilities of grey parrots. *The Analysis of Verbal Behavior* 20: 141–151

38. Aiello, L., and P. Wheeler (1995). The Expensive-Tissue Hypothesis: The brain and the digestive system in human

and primate evolution. *Current Anthropology*, 36(2): 199-221

39. Rolian, C., Lieberman, D.E. and B. Hallgrímsson (2010). The coevolution of human hands and feet. *Evolution* 64: 1558-1568.

40. Nagel, M., Jansen, P.R., Stringer, S., Watanabe, K., de Leeuw, C.A. et al. (2018). Meta-analysis of genome-wide association studies for neuroticism in 449,484 individuals identifies novel genetic loci and pathways. *Nature Genetics* 50: 920–927

41. Stone, E. (2017). The Voyagers. Nature Astronomy 1(896)

42. Fidler, A. E., van Oers, K., Drent, P. J., Kuhn, S., Mueller, J. C., and B. Kempenaers (2007). Drd4 gene polymorphisms are associated with personality variation in a passerine bird. *Proceedings. Biological Sciences* 274(1619): 1685–1691

43. Thomson, C.J., Hanna, C.W., Carlson, S.R., Rupert, J.L. (2013). The -521 C/T variant in the dopamine-4-receptor gene (DRD4) is associated with skiing and snowboarding behavior. *Scandinavian Journal of Medicine & Science Sports* 23(2): e108-e113

44. Kourt, A. (2017). *The curiosity gene: On the origin of humankind by means of intrinsic motivation*. US: CreateSpace Independent Publishing Platform.

45. Byrne, R.W. (2013). Animal curiosity. *Current Biology* 23(1): R469-R470

46. Gräslund, B. (2005). *Early humans and their world*. UK: Routledge Publishing.

47. Wesołowska, W. and T. Wesołowski (2014). L. eucochloridium sporocysts and snail host behaviour. *Journal of Zoology* 292: 151-155

48. Riley, D. R., Sieber, K. B., Robinson, K. M., White, J. R., Ganesan, A., Nourbakhsh, S., J.C. Dunning Hotopp (2013). Bacteria-human somatic cell lateral gene transfer is enriched in cancer samples. *PLoS Computational Biology* 9(6): e1003107

49. Ryan, F. (2009). *Virolution*. London, UK: Collins Publishers

50. Poljak, N., Klindzic, D. and M. Kruljac (2019). Effects of exoplanetary gravity on human locomotion ability. *The Physics Teacher* 57(6): 378–381

51. Griffiths, A.J.F., Miller, J.H., Suzuki, D.T. et al. (2000). *An introduction to genetic analysis 7th edition.* New York: W. H. Freeman.

52. Machado, J. M. Lopes-Lima (2011). Calcification mechanism in freshwater mussels: Potential targets for cadmium. *Toxicological and Environmental Chemistry* 93: 10.1080/02772248.2010.503656

53. Chandrasekaran, C. and E. Betrán (2008). Origins of new genes and pseudogenes. *Nature Education* 1(1): 181

54. Hoffmann, F. G., Opazo, J. C., and J.F. Storz (2010). Gene cooption and convergent evolution of oxygen transport hemoglobins in jawed and jawless vertebrates. *Proceedings of the National Academy of Sciences of the United States of America* 107(32): 14274–14279

55. Tautz, D., Neme, R. and T. Domazet-Lošo (2013). Evolutionary origin of orphan genes. In *eLS* (Ed.) DOI:10.1002/9780470015902.a0024601

56. Schlötterer, C. (2015). Genes from scratch - The evolutionary fate of *de novo* genes. *Trends in Genetics* 31(4): 215–219

57. Suzuki, M., Saruwatari, K., Kogure, T. et al. (2009). An acidic matrix protein, *Pif*, is a key macromolecule for nacre formation. *Science*. 325(5946): 1388-1390.

58. Zhang, R., Qin, M., Shi, J., Tan, L., Xu, J., Tian, Z., Wu, Y., Li, Y., Li, Y. and N. Wang (2018). Molecular cloning and characterization of *Pif* gene from pearl mussel, *Hyriopsis cumingii*, and the gene expression analysis during pearl formation. *3 Biotech* 8(4): 214

59. Suzuki, M., Iwashima, A., Kimura, M., Kogure, T. and H. Nagasawa (2013). The molecular evolution of the *pif* family proteins in various species of mollusks. *Marine Biotechnology* 15(2): 145-158.

60. De Plaa, J., Werner, N., Bleeker, J.A.M., Vink, J. and J.S. Kaastra (2007). Constraining supernova models using the hot gas in clusters galaxies. *Astronomy & Astrophysics* 465(345)

61. Perets, H.B., Gal-Yam, A., Mazzali, P.A., Arnett, D., Kagan, D. et al. (2010). A faint type of supernova from a white dwarf with a helium-rich companion. *Nature* 465(7296): 322

62. NASA (2017). *Got calcium?* NASA Goddard Space Flight Center Educator's Corner. https://imagine.gsfc.nasa.gov/educators/calcium/got_calcium_litho.html

63. Nakabachi, A., Yamashita, A., Toh, H., Ishikawa, H., Dunbar, H.E., Moran, N.A., Hattori, M. (2006). The 160-Kilobase genome of the bacterial endosymbiont Carsonella. *Science* 314(5797): 267-267

64. Sebastian, R. Schmidl, F., Ekness, K., Sofjan, K. et al. (2019). Rewiring bacterial two-component systems by modular DNA-binding domain swapping. *Nature Chemical Biology* DOI: 10.1038/s41589-019-0286-6

65. Christensen, C. B., Lauridsen, H., Christensen-Dalsgaard, J. Pedersen, M. and P.T. Madsen (2015). Better than fish on land? Hearing across metamorphosis in salamanders. *Proceedings of the Royal Society B: Biological Sciences* 282(1802): 20141943

66. Niimura, Y. (2012). Olfactory receptor multigene family in vertebrates: from the viewpoint of evolutionary genomics. *Current Genomics* 13(2): 103–114.

67. Watanabe, H., Fujisawa, T. and T.W. Holstein (2009). Cnidarians and the evolutionary origin of the nervous system. *Development, Growth & Differentiation* 51: 167-183

68. Crisp, A., Boschetti, C., Perry, M., Tunnacliffe, A., and G. Micklem (2015). Expression of multiple horizontally acquired genes is a hallmark of both vertebrate and invertebrate genomes. *Genome Biology* 16(1): 50

69. Salzberg, S. L. (2017). Horizontal gene transfer is not a hallmark of the human genome. *Genome Biology* 18(1), 85

70. Heidmann, O., Béguin, A., Paternina, J., Berthier, R., Deloger, M., Bawa, O. and T. Heidmann (2017). HEMO, an ancestral endogenous retroviral envelope protein shed in the blood of pregnant women and expressed in pluripotent stem cells and tumors. *Proc Natl Acad Sci U S A.* 114(32): E6642-E6651

71. Peddu, V., Dubuc, I., Gravel, A., Xie, H., Huang, M.-L., Tenenbaum, D., Jerome, K.R. et al. (2019). Inherited chromosomally integrated human herpesvirus 6 demonstrates tissue-specific RNA expression in vivo that

correlates with an increased antibody immune response. *Journal of Virology* 94(1): e01418-19

72. Aparicio, S., Chapman, J., Stupka, E., Putnam, N., Chia, J.M. et al. (2002). Whole-genome shotgun assembly and analysis of the genome of Fugu rubripes. *Science* 297(5585): 1301-1310

73. Koshiba-Takeuchi, K., Mori, A. D., Kaynak, B. L., Cebra-Thomas, J., Sukonnik, T., Georges, R. O., Latham, S., Beck, L., Henkelman, R. M., Black, B. L., Olson, E. N., Wade, J., Takeuchi, J. K., Nemer, M., Gilbert, S. F. and B.G. Bruneau (2009). Reptilian heart development and the molecular basis of cardiac chamber evolution. *Nature* 461(7260): 95–98

74. Roger, A.J., Muñoz-Gómez, S.A., Kamikawa, R. (2017). The origin and diversification of mitochondria. *Current Biology* 27(21): R1177-R1192

75. Naumann, R. K., Ondracek, J. M., Reiter, S., Shein-Idelson, M., Tosches, M. A., Yamawaki, T. M., and G. Laurent (2015). The reptilian brain. *Current Biology* 25(8): R317–R321

76. Reiner, A. (1990). The triune brain in evolution. Role in paleocerebral functions. Paul D. MacLean. Plenum, New York, 1990. xxiv, 672 pp., illus. $75. *Science* 250(4978): 303-305

77. Tatsumi, N., Kobayashi, R., Yano, T., Noda, M., Fujimura, K., Okada, N. and M. Okabe. (2016). Molecular developmental mechanism in polypterid fish provides insight into the origin of vertebrate lungs. *Scientific Reports* 6: 30580

78. Li, J.Z., Bunney, B.G., Meng, F., Hagenauer, M.H., Walsh, D.M. et al. (2013). Circadian patterns of gene expression in the human brain and disruption in major depressive

disorder. *Proceedings of the National Academy of Sciences* 110 (24) 9950-9955

79. Goldinger, A., Shakhbazov, K., Henders, A. K., McRae, A. F., Montgomery, G. W., and J.E. Powell (2015). Seasonal effects on gene expression. *PloS One* 10(5): e0126995

80. Cohen, F.S. (2016). How Viruses Invade Cells. *Biophysical Journal* 110(5): 1028–1032

81. Brüssow, H. (2009). The not so universal tree of life or the place of viruses in the living world. *Philosophical transactions of the Royal Society of London. Series B, Biological Sciences* 364(1527): 2263–2274

82. Forterre, P. (2015). The universal tree of life: an update. *Frontiers in Microbiology* 6(717)

83. Lwoff, A. (1957). The concept of virus. *Journal of General Microbiology* 17: 239–53

84. Randich, A. M. and Y.V. Brun (2015). Molecular mechanisms for the evolution of bacterial morphologies and growth modes. *Frontiers in Microbiology* 6: 580

85. Herskowitz, I. (1988). Life cycle of the budding yeast Saccharomyces cerevisiae. *Microbiological Reviews* 52(4): 536–553

86. Farrell, K.W. (1976). Flagellar regeneration in *Chlamydomonas reinhardtii*: Evidence that cycloheximide pulses induce a delay in morphogenesis. *Journal of Cell Science* 20: 639-654

87. Holmes E. C. (2011). What does virus evolution tell us about virus origins? *Journal of Virology* 85(11): 5247–5251

88. Moreira, D and P. López-García (2009). Ten reasons to exclude viruses from the tree of life. *Nature Reviews Microbiology* 7(4): 306-311

89. Domingo, E. (2020). Introduction to virus origins and their role in biological evolution. *Virus as Populations* 1–33

90. Koonin E. V. (2009). On the origin of cells and viruses: Primordial virus world scenario. *Annals of the New York Academy of Sciences* 1178(1): 47–64

91. Wessner, D. R. (2010) The origins of viruses. *Nature Education* 3(9): 37

92. Durzyńska, J. and A. Goździcka-Józefiak, (2015). Viruses and cells intertwined since the dawn of evolution. *Virology Journal* 12(169)

93. Krupovic, M. (2012). Recombination between RNA viruses and plasmids might have played a central role in the origin and evolution of small DNA viruses. *BioeEssays* 34: 867-870

94. Vogan, A.A. and P.G. Higgs (2011). The advantages and disadvantages of horizontal gene transfer and the emergence of the first species. *Biology Direct* 6(1)

95. Suttle, C.A. (2007). Marine viruses - major players in the global ecosystem. *Nature Reviews Microbiology* 5: 801-12

96. Weit, J.S. and S.W. Wilhelm (2013, June 30). *An ocean of viruses.* The Scientist. https://www.the-scientist.com/features/an-ocean-of-viruses-39112

97. Garwood, R.J. and G.D. Edgecombe (2011). Early terrestrial animals, evolution, and uncertainty. *Evolution: Education and Outreach* 4: 489–501

98. Palazzo, A.F. and T.R. Gregory (2014). The case for junk DNA. *PLoS Genetics* 10(5): e1004351

99. ENCODE Project Consortium (2012). An integrated encyclopedia of DNA elements in the human genome. *Nature* 489(7414): 57–74

100. Doolittle, W. (2013). Is junk DNA bunk? A critique of encode. *Proceedings of the National Academy of Sciences of the United States of America* 110(14): 5294-5300

101. Graur, D., Zheng, Y., Price, N., Azevedo, R.B.R., Zufall, R.A. and E. Elhaik (2013). On the immortality of television sets: "Function" in the human genome according to the evolution-free gospel of ENCODE. *Genome Biology and Evolution* 5 (3): 578–590

102. Jagannathan, M., Cummings, R. and Y.M. Yamashita. (2018). A conserved function for pericentromeric satellite DNA. *eLife* DOI: 10.7554/eLife.34122

103. Baalsrud, H.T., Tørresen, O.K., Solbakken, M.H., Salzburger, W., Hanel, R. et al. (2018). De Novo gene evolution of antifreeze glycoproteins in codfishes revealed by whole genome sequence data. *Molecular Biology and Evolution* 35(3): 593–606

104. Lorenz, D.M., Jeng, A., and M.W. Deem (2011). The emergence of modularity in biological systems. *Physics of Life Reviews* 8(2): 129-60

105. Serban, M. (2020). Exploring modularity in biological networks. *Philosophical transactions of the Royal Society of London. Series B, Biological Sciences* 375(1796): 20190316

106. Porcar, M., Latorre, A. and A. Moya (2013). What symbionts teach us about modularity. *Frontiers in Bioengineering and Biotechnology* 1(14)

107. Espinosa-Soto, C. (2018). On the role of sparseness in the evolution of modularity in gene regulatory networks. *PLoS Computational Biology* 14(5): e1006172

108. Vinicius, L. (2010). *Modular evolution: How natural selection produces biological complexity.* UK: Cambridge University Press

109. Gould, S. J. (1989). *Wonderful life: The Burgess Shale and the nature of history.* New York: Norton

110. Blount, Z.D., Borland, C. Z. and R.E. Lenski (2008). Historical contingency and the evolution of a key innovation in an experimental population of *Escherichia coli*. *Proceedings of the National Academy of Sciences* 105(23): 7899–906

111. Dobler, S., Dalla, S., Wagschal, V. and A.A. Agrawal (2012). Community-wide convergent evolution in insect adaptation to toxic cardenolides by substitutions in the Na,K-ATPase. *Proceedings of the National Academy of Sciences of the United States of America* 109(32): 13040–13045

112. Dawkins, R. (1976). *The selfish gene.* UK: Oxford University Press

113. Ward, P. (2012, November 1). *What may become of Homo sapiens.* Scientific American. https://www.scientificamerican.com/article/what-may-become-of-homo-sapiens/

www.ingramcontent.com/pod-product-compliance
Lightning Source LLC
Chambersburg PA
CBHW050002230526
45465CB00003BB/1219